Martin Schaffer

Key Aspects of Random Number Generation

Martin Schaffer

Key Aspects of Random Number Generation

Uniqueness, Efficiency, Privacy

VDM Verlag Dr. Müller

Imprint

Bibliographic information by the German National Library: The German National Library lists this publication at the German National Bibliography; detailed bibliographic information is available on the Internet at http://dnb.d-nb.de.

Any brand names and product names mentioned in this book are subject to trademark, brand or patent protection and are trademarks or registered trademarks of their respective holders. The use of brand names, product names, common names, trade names, product descriptions etc. even without a particular marking in this works is in no way to be construed to mean that such names may be regarded as unrestricted in respect of trademark and brand protection legislation and could thus be used by anyone.

Cover image: www.purestockx.com

Publisher:
VDM Verlag Dr. Müller Aktiengesellschaft & Co. KG , Dudweiler Landstr. 125 a, 66123 Saarbrücken, Germany,
Phone +49 681 9100-698, Fax +49 681 9100-988,
Email: info@vdm-verlag.de

Zugl.: Klagenfurt, Universität, Diss., 2007

Produced in USA and UK by:
Lightning Source Inc., La Vergne, Tennessee, USA
Lightning Source UK Ltd., Milton Keynes, UK
BookSurge LLC, 5341 Dorchester Road, Suite 16, North Charleston, SC 29418, USA

ISBN: 978-3-8364-9482-3

To my family.

One may not mistake the things,
which seem to be unlikely and unnatural,
for something which is absolutely impossible.

CARL FRIEDRICH GAUSS (1777 − 1855)

Abstract

An intrinsic requirement of information and communication systems is the unique
identification of objects and entities. Typical examples include telephone numbers,
email addresses, MAC-addresses, domain names and, in the context of smartcards,
Integrated ChipCard Serial Numbers. Complex IT-systems generally consist of several
software and hardware components, each of which must be designed to be sufficiently
robust. In practice, it is intrinsic that these components are uniquely identifiable
during operations. If this is not considered, two distinct components might share the
same identifier, which could lead to reduced operating ability or even system failure.

One solution for this problem is to use locally generated system-wide unique identifiers. The requirement of *system-wide* uniqueness is necessary, since in general several
instances are involved in the implementation process who cannot necessarily interact
with each other. For efficiency reasons the generation process should work without
interactions and with as little computations and space as possible.

Most of the approaches for the generation of identifiers currently used in practice work
efficiently, however, they fulfill the requirement of system-wide uniqueness only with
a high probability and not entirely. The reason for this lies in the fact that these
approaches are also used for other areas where the protection of the issuer's privacy
is an additional requirement. For instance, that one cannot efficiently find out *who*
generated a given identifier (e.g. if it is used as a pseudonym).

Algorithms which ensure system-wide uniqueness of identifiers can also be useful for
cryptographic purposes. If, for instance, two entities were accidentally provided with
the same signature key, then it might be impossible to decide, which of the two generated a given signature. Another typical example concerns the remote central locking of
cars. If the transmitted key is not guaranteed to be unique then it might happen that
one accidentally opens a different car than intended. Again, the specific generation
processes can be used to ensure system-wide uniqueness. However, one may only use
solutions which provide a sufficient quality of randomness in the output.

Some IT-systems require the existence of *one* single trusted party to fulfill critical
tasks (e.g. the generation of signature keys). However, in the real world this is an
unrealistic assumption because no trustworthy instance exists. One might think of

political elections, for instance, where the tally cannot be carried out by a single party. Hence, there is a need to provide solutions which can share critical tasks among a set of instances in a fair way. In the cryptographic context the techniques of *secure multi-party computation* are useful to give practical solutions in this area. Amongst others, a widely used technique is the distributed generation of secret keys, which are then used in shared form (e.g. for the shared decryption of an election result). One yet unsolved problem in this context is the distributed generation of keys which, if reconstructed, are guaranteed to be *system-wide unique* whilst providing a certain quality of randomness.

This work uses several examples to motivate why the mentioned generation processes are necessary. Furthermore, related schemes are analyzed with respect to several stated requirements. Novel solutions are presented on an abstract level and practical examples as well as applications are discussed. Furthermore, widely known techniques are used as building blocks for the design of multi-party protocols that can be used to generate system-wide unique secret keys for discrete-logarithm-based cryptosystems. To provide the same level of (communication) efficiency as the solutions currently used in practice, a special algebraic structure is presented, called *fusion*. Accordingly, variants of the Discrete Logarithm Problem, the Diffie-Hellman Problem and the Decision Diffie-Hellman Problem are defined. The first two are shown to be computationally equivalent to their ordinary variants. Solving the third variant leads to an efficient solution of the associated ordinary variant. Since the presented algebra also provides some other advantages it is described separately. The final part of this work presents two authentication schemes which protect the privacy of the user, whilst preserving the interests of the verifier. Privacy in this context means that the user can remain anonymous unless he or she behaves dishonestly. As soon as the user behaves dishonestly *after* an authentication, his or her real identity can be disclosed by a set of trusted parties using secure multi-party computation.

Acknowledgements

Firstly, I would like to heartfully thank Professor Patrick Horster for his useful comments, his support in general and especially for giving me a lot of freedom in my research activities which cannot be taken for granted these days. Secondly, my thanks go to Professor Winfried Müller for his feedback in general and especially concerning some algebraic and number-theoretic aspects.

Very special thanks go to my two friends and colleagues Peter Schartner and Stefan Rass for their support and co-work on several topics. Without their help this work would never have been of such quality. Big thanks also go to my dear friend Dieter Sommer who was always willing to help whenever I had problems.

My thanks concerning proof-reading and mental support go to Arthur Pitman, David Herrgesell, Dagmar Cechak, Birgit Winkler and my family.

Finally, I would like to thank my wife Sandra for her endless patience and support.

TABLE OF CONTENTS

4   The Concept of Fusion   **67**

INTRODUCTION

Chapter 1

1.1 Motivation

To provide flexibility, modern IT-systems generally consist of several self-contained software components, each of which must be designed carefully if the system is to be robust. Additionally, robustness implies that the software components interact with each other correctly and thus are uniquely identifiable. For instance, consider an object-oriented programming language. If two objects used by the same software accidentally share the same identifier, a system failure might occur if the wrong one is accessed. This in turn might cause unnecessary delays or failure in the associated application. For instance, a transportation system might cause traffic jams due to switching to an emergency program. Another example is a factory, where the production might be interrupted for some time, delaying the delivery of a product. One solution to this problem is to at least guarantee that no two objects within the same software system share the same identifier. Since in general companies do not implement all components of a software on their own, there is a need for algorithms by which it is possible to locally generate (provably) system-wide unique identifiers or in a more general context, system-wide unique numbers. We call such a generation process *collision-free number generation*. One efficient approach is to provide each entity of a system with a system-wide unique identifier, concatenated to a counter which is incremented every time the generation process is run. As long as incrementing the counter gives no overflow, the output is system-wide unique. Such a solution could provide sufficient robustness (regarding safety) for the system.

Unfortunately, uniqueness is often not the only requirement. In several situations, it is important that one is not able to efficiently find out *who* generated a given number. A typical example for such a situation is anonymous authentication based on pseudonyms. The user wants to prove that he or she is a registered user (i.e. is the owner of a registered pseudonym), but does not want to reveal any personal information. At the same time the user needs to be uniquely identifiable through the pseudonym by a revocation center in case of misbehavior. A further requirement might be that it is infeasible to efficiently decide if a set of pseudonyms corresponds to the same holder. Being able to make such a decision can already decrease the degree of privacy protection. For instance, one may use several pseudonyms for different electronic transactions, so that his or her activities are not traceable. If it is possible to efficiently identify which pseudonyms belong to the same holder, then the use of several pseudonyms ceases to make sense. Thus, beside system-wide uniqueness, there is sometimes the additional need to protect the privacy of the issuer of a unique number, i.e. hiding his or her identity and ensuring that the issued numbers cannot be traced.

The standard for the generation of Universally Unique Identifiers [IT04] provides algorithms which either ensure system-wide uniqueness of the generated identifiers *or* protect the privacy of the corresponding issuer, but not both at the same time. Since privacy has become more and more important in the recent years, many systems only use the algorithms proposed in [IT04] where strong privacy is provided. The accidental double-issuing of the same identifier is accepted to occur with a low probability.

The problem, however, is that the probability theory often plays dirty tricks. In the ideal case, where events are truly random and independent, events with negligible probability occur rarely. However, this is an impractical assumption, since true randomness is hard to achieve artificially. For instance, if one chooses identifiers by using an algorithm which uses a bad or defective pseudo-random number generator as a subroutine, then a collision may occur sooner than expected. Moreover, an event may never happen or may occur immediately. A remarkable example of unexpected collisions occurred in two lotteries. In 1977, 205 gamblers in Germany won by playing the same numbers that were drawn in Holland the previous week [Wal98]. Apart from the fact that so many people had the idea of playing the same numbers, the probability that two different countries drawing the same numbers within one week is exceptionally low. This historical event is an excellent example that one should not just rely on the safety (in particular robustness) of a system with the hope that an event happens rarely. It might happen sooner than one thinks.

Besides avoiding system failures, collision-free number generation is also useful for cryptographic purposes. For instance, if two entities accidentally choose the same secret key, then they may be able to mutually impersonate each other. As long as such a collision is not publicly detectable, there seems to be no problem. Unfortunately, in several security applications and cryptosystems a collision is easily detectable, allowing attackers to take advantage of it. The following sketches some examples which are discussed in more detail later on in Chapter 2.

Car Keys: Nowadays, the majority of the cars have remote central locking. Code Hopping Encoders are widely used to authenticate a given remote control to the corresponding car [Kee03]. Both the car and the remote control use a sequence of one-time keys generated by using a pseudo-random number generator initialized with the same seed. Synchronization is ensured by some kind of sliding-window technique. Notice that common pseudo-random number generators can normally at most avoid local collisions, but are not designed to avoid world-wide ones. As a consequence it might happen that one accidentally opens a different car than intended.

Decryption and Signature Keys: If the modulus of an RSA signature key [RSA78] is not unique, which can be the case if the associated prime-factors are chosen at random, it might happen that two different users share the same modulus. In the recent years, digital signatures became an accepted alternative to hand-written signatures (in some countries even by law). Consider the case where a signature is involved in a court case (e.g. tax fraud) and is necessary to prove one's innocence. If the court is not able to decide, whether the accused or another person, who by accident has the same signature key, signed a particular document, then the signature might not be admitted as evidence. Hence the guilty person might be released and the innocent sentenced to stay in prison. The same problem arises for ElGamal-like signature schemes [ElG85a, NIS94]. In case of RSA and ElGamal-like [ElG85a, CS98] encryption, a collision enables unauthorized decryptions.

Probabilistic Encryption and Signature Schemes: Notice that identical keys are not the only problem. Several encryption and signature schemes are probabilistic, which means that some additional fresh randomness (called randomizer) is involved in the generation process of a ciphertext or a signature. If one signs different messages with ElGamal-like signature schemes and accidentally chooses the same randomizer, then an eavesdropper can extract the secret signature key from the protocol transcript. In case of ElGamal-like encryption schemes, an eavesdropper might obtain some information about the plaintext. This is generally not as problematic as the disclosure of the secret key, but can be fatal anyhow if the plaintext itself is a key.

Proofs of Knowledge: Another example is the use of interactive proofs of knowledge [GMR89, BG93], in particular Σ-protocols [Cra96]. In such protocols, a prover proves in zero-knowledge the knowledge of a secret to a verifier. Zero-knowledge hereby means that the verifier is convinced that the prover knows the secret, but obtains no additional information. Sending the same first-message twice and then responding to two different challenges may enable the verifier to extract the secret. This property is necessary to prove that the interactive proof is designed correctly, i.e. that it is a proof of knowledge, but this must not happen in practice. Since the generation process of the first-message generally involves random numbers, a collision can occur even though this generally happens only with low probability.

1.2   Contribution of this Work

The contribution of this work spans several areas. The initial contribution is the introduction of the concept of *collision-free number generation* on an abstract level. Thereby, we are the first to define explicit requirements which cover a large range of applications. Besides ensuring that keys are unique (the focus of the majority of related schemes), this concept can also be used to improve the safety of the systems. Practical implementations are given, some of which better fulfil the formulated requirements, compared to related approaches. The emphases of this book are listed below.

Efficient Constructions: The presented techniques are developed to be relevant in practice. Hence, throughout the work, particular care is taken to be efficient in terms of communication, time and space.

Privacy Issues: Our approach to collision-free number generation has a special focus on preserving the privacy of the generating party. Additionally, an authentication scheme is presented, which preserves the privacy of authenticators.

Cryptographic Aspects (Optimizations): An additional, however widely independent, contribution is the invention of the *concept of fusion*. First designed as a simplified approach towards distributed generation of unique keys for discrete-logarithm-based threshold cryptosystems, this concept turned out to generally speed up cryptographic applications in particular situations. For example, if one uses a collision-free number generator whose minimal output-length is more than necessary from security point of view, then using this concept can bring an enormous speed-up. Besides introducing the concept of fusion on an abstract level, we prove its correctness and define novel kinds of the Discrete Logarithm Problem [McC90], the Diffie-Hellman Problem [DH76] and the Decision Diffie-Hellman Problem [Bon98]. Furthermore, we prove that the first two are computationally equivalent to their ordinary variants, whereas the latter is at least as hard to solve as in the ordinary setting. Showing the converse direction is currently an open problem which yields an interesting conjecture: if the computationally equivalence of the two problems cannot be shown, then the new one seems to be a stronger problem than the ordinary one. Thus, if the Decision Diffie-Hellman Problem is efficiently solved directly (i.e. without solving the Discrete Logarithm Problem or the Diffie-Hellman Problem), then related cryptosystems like the ElGamal cryptosystem [ElG85a] or the Cramer-Shoup cryptosystem [CS98] will become vulnerable. However, if our conjecture remains unrefuted, then such cryptosystems will still remain secure within the fusion-setting.

Cryptographic Aspects (Threshold Schemes): The basic design of collision-free number generation focuses on the single-server setting, where each generator is run by

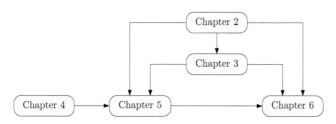

Fig. 1.1: Dependencies between the Core-Chapters.

its holder which is actually *one* instance. If, however, this concept is used for generating secret keys for threshold cryptosystems [DF90], then fairly advanced techniques are necessary. The challenge is that a set of instances is jointly responsible for generating shares of a secret without reconstructing it. Several solutions for this problem exist, but none of them guarantees that the (virtual) secret, if reconstructed, is system-wide unique. The difficulty is, that during the generation process the processors are not allowed to reconstruct the secret or any partial information. In this work, we present three solutions for this problem for discrete-logarithm-based threshold cryptosystems, first on an abstract level and then with respect to practical implementations. However, only one of these approaches seems to be of practical relevance, since it is the only one whose security can be reduced to an unsolved computational problem. Moreover, it is as (communication) efficient as already existing distributed key generation protocols.

Several applications are presented in each chapter. Additionally, a larger application scenario is given, where several of the introduced techniques are applied. We present two unlinkable anonymous authentication schemes with optional shared anonymity revocation and linkability. Anonymity we achieve through using pseudonyms, which are guaranteed to be system-wide unique with the help of collision-free number generation. Unlinkability and optional revocation are provided through elaborate discrete-logarithm-based techniques with respect to suitable (algebraic) groups of prime order.

1.3 Organization

The dependencies between the chapters are illustrated in Figure 1.1. An edge shows the recommended path, a reader should follow when examining this work. In the following, we sketch the content of each chapter.

Chapter 2: Initially, some examples are presented where collisions among numbers cause some vulnerabilities. These cover Universally Unique Identifiers and pseudonym systems (i.e. safety) on the one hand. On the other hand, they cover private and public key cryptosystems, probabilistic schemes and Σ-protocols (i.e. security).

Chapter 3: Based on the vulnerabilities identified in Chapter 2, several requirements are defined, under which related approaches towards collision-free number generation are analyzed on an abstract level. The basic construction principle is presented and three types of collision-free number generators are defined. Furthermore, some practical implementations are sketched and analyzed. Finally, some applications are given. Preliminary ideas of this chapter can be found in [SS05, SSR07a, SSR07b] and [SS08].

Chapter 4: The concept of fusion is introduced and proven to have the desired properties. Furthermore, the aforementioned computational problems are defined and reduced to several unsolved existing problems. Additionally, it is shown that in special situations this new concept brings a speed-up for discrete-logarithm-based cryptosystems. Preliminary ideas of this chapter can be found in [RS06, SR07] and [SR08].

Chapter 5: Three techniques are presented in this chapter, to securely share a simple collision-free number generator among a set of n processors of which $t < n/2$ are allowed to actively behave dishonestly. Hereby, the concept of fusion gives the only provably secure protocol, while at the same time being as (communication) efficient as the distributed key generation protocols currently used in practice. Preliminary ideas of this chapter have been given in [RSS06b, SR07] and [SR08].

Chapter 6: Finally, two authentication schemes are presented. A preliminary version of the first one can be found in [SS06]. The second one has similar properties, but is far more efficient and could even be used on power-limited devices. A preliminary version of the second approach with an emphasis of casino-games can be found in [RSS06a].

The work closes with some remarks on future prospects.

Remark: A previous version of this book has been published as a PhD-thesis [Sch07].

1.4 Notation, Preliminaries, and Further Remarks

Appendix A contains a list of symbols and notations used in this work. A list of acronyms can be found in Appendix B. The structure of this work allows to easily distinguish between preliminary and novel results. Preliminary results are essentially located in sections entitled *Preliminaries* or *Related Work*. Such sections generally contain building blocks that are necessary to understand our approaches.

Throughout this work, we assume that the reader is familiar with the basics of security, cryptography [MVO96, Sch96, PHS03, HMV04], number theory [Bun02] and algebra [SW78, DM84, Bos05]. In the context of residue classes we mostly do not write MOD n for computations modulo n, since this will always be clear from the context.

Throughout this work the male gender is used exclusively to provide a fluent reading.

VULNERABLE APPLICATIONS

Chapter 2

2.1 Introduction

In this chapter, selected applications are presented where collisions among numbers or identifiers might lead to security holes or system failures. Discussions on Universally Unique Identifiers (Section 2.3) and pseudonyms (Section 2.4) are given on an abstract level, which is sufficient since the vulnerabilities can be identified easily without giving details. In the context of key generation (Section 2.5), probabilistic encryption (Section 2.6), probabilistic signature generation (Section 2.7) and interactive proofs of knowledge (Section 2.8), the occurrence of security holes strongly depends on the considered scheme. Thus, some particular cryptosystems are analyzed in more detail.

2.2 Preliminaries

The following discussion outlines basic public key cryptosystems. For consistency, the discrete-logarithm-based schemes are described for a cyclic group $\mathbb{G}_q$ of prime order q.

2.2.1 The ElGamal Encryption Scheme

In [ElG85a, ElG85b], an encryption scheme has been proposed which is based on the Diffie-Hellman Problem [DH76]. ElGamal encryption with respect to $\mathbb{G}_q$ works as follows. During the key generation process the secret key z is chosen at random from $\mathbb{Z}_q$. The corresponding public key is defined as $h = g^z$, where g is a generator of $\mathbb{G}_q$. A plaintext $m \in \mathbb{G}_q$ is encrypted as follows, with respect to h:

$$u = g^r, \quad e = mh^r, \quad r \in_R \mathbb{Z}_q$$

where (u, e) denotes the ciphertext. Given the secret key z and (u, e), the plaintext can be obtained through the following computation:

$$m = e(u^z)^{-1}$$

The ElGamal encryption scheme is insecure against a chosen-ciphertext attack [Sho98]: setting $e' = eb$, where $b \in \mathbb{G}_q$, the decryption gives mb. If a cryptosystem has such a property it does not provide *non-malleability* [DDN91], i.e. one can modify the ciphertext in such a way that the resulting plaintext is meaningfully related to the original one. In practice, whether this attack is feasible or not and would bring a real advantage to the adversary depends on the application, i.e. the scheme can be secure enough. Nonetheless, a modification has been proposed in [CS98] which withstands chosen-ciphertext attacks and is given in the following section.

2.2.2 The Cramer-Shoup Encryption Scheme

Let g_1 and g_2 be two distinct generators of $\mathbb{G}_q$, where the mutual discrete logarithms are unknown. The Cramer-Shoup encryption scheme [CS98] works as follows. The secret key consists of the six components x_1, x_2, y_1, y_2, z_1 and z_2, chosen at random from $\mathbb{Z}_q$. The public key consists of the three components c, d and h, defined as follows:

$$c = g_1^{x_1} g_2^{x_2}, \quad d = g_1^{y_1} g_2^{y_2}, \quad h = g_1^{z_1} g_2^{z_2}$$

The encryption of a plaintext $m \in \mathbb{G}_q$ works in a similar way to ElGamal encryption:

$$u_1 = g_1^r, \quad u_2 = g_2^r, \quad e = mh^r, \quad r \in_R \mathbb{Z}_q$$

Additionally, a verification basis $v \in \mathbb{G}_q$ for the ciphertext is computed as follows:

$$v = c^r d^{r\mathcal{H}(u_1, u_2, e)}$$

where $\mathcal{H}$ is a cryptographic hash function and (u_1, u_2, e, v) denotes the ciphertext. Given the secret key $(x_1, x_2, y_1, y_2, z_1, z_2)$ and (u_1, u_2, e, v), one has to check if the encryption has been performed correctly, i.e. (u_1, u_2, e, v) is truly the output of the encryption algorithm with respect to the public key (c, d, h):

$$v \stackrel{?}{=} u_1^{x_1} u_2^{x_2} (u_1^{y_1} u_2^{y_2})^{\mathcal{H}(u_1, u_2, e)}$$

If the check fails, decryption is refused, otherwise m is computed straightforwardly:

$$m = e(u_1^{z_1} u_2^{z_2})^{-1}$$

2.2.3 The Simplified Cramer-Shoup Encryption Scheme

In [CS98, CS04], a simplified and hence more efficient version of the Cramer-Shoup cryptosystem has been presented, which works as follows. During the key generation process, the four secret key components w, x, y and z are chosen at random from $\mathbb{Z}_q$. The public key then consists of the four components g_2, c, d and h, defined as follows:

$$g_2 = g_1^w, \quad c = g_1^x, \quad d = g_1^y, \quad h = g_1^z$$

Notice that g_2 is not a system parameter anymore. The encryption process, including the generation of the verification basis, is done exactly as in the basic scheme. Prior to the decryption of a message, given the secret key (w, x, y, z) and (u_1, u_2, e, v), one has to perform the following two checks:

$$u_2 \stackrel{?}{=} u_1^w, \quad v \stackrel{?}{=} u_1^{x+y\mathcal{H}(u_1, u_2, e)}$$

If either of the verifications fails, the decryption process is rejected. Otherwise, m is obtained through the standard ElGamal decryption process.

2.2.4 The ElGamal Signature Scheme

In [ElG85a, ElG85b], a signature scheme has been proposed in addition to the encryption scheme. With respect to $\mathbb{G}_q$ it works as follows. The public key is defined as $y = g^x$, where $x \in_R \mathbb{Z}_q$ is the corresponding secret key. A signature $(R, s) \in \mathbb{G}_q \times \mathbb{Z}_q$ for a document $M \in \{0, 1\}^*$ is generated as follows:

$$R = g^r, \quad s = r^{-1}(\mathcal{H}(M) - xf(R)), \quad r \in_R \mathbb{Z}_q$$

where $\mathcal{H} : \{0, 1\}^* \to \mathbb{Z}_q$ is a cryptographic hash function and $f : \mathbb{G}_q \to \mathbb{Z}_q$ a deterministic one-way function. If $\mathbb{G}_q$ is a subgroup of $\mathbb{Z}_p^*$, one can set $f(R) := R \text{ MOD } q$, for instance. Another solution is to use a cryptographic hash function for f [BPVY00]. The verification of (R, s) with respect to M and y is performed in the following manner:

$$g^{\mathcal{H}(M)} \stackrel{?}{=} y^{f(R)} R^s$$

2.2.5 The Schnorr Signature Scheme

In the Schnorr signature scheme [Sch89], the public and the secret key are defined similarly to those in the ElGamal signature scheme. A signature $(c, s) \in \mathbb{Z}_q \times \mathbb{Z}_q$ with respect to a document $M \in \{0, 1\}^*$ and the secret key x is generated as follows:

$$c = \mathcal{H}(M \| g^r), \quad s = r - cx, \quad r \in_R \mathbb{Z}_q$$

where $\mathcal{H} : \{0, 1\}^* \to \mathbb{Z}_q$ is a cryptographic hash function. Notice that this variant slightly differs from [Sch89]. Given (c, s), M and y, the verification process is:

$$c \stackrel{?}{=} \mathcal{H}(M \| y^c g^s)$$

Notice that although this signature scheme is an interactive proof for the knowledge of a discrete logarithm, it has been made non-interactive by using the Fiat-Shamir paradigm [FS87]. Further information on this topic is given in Section 5.2.3.1.

2.2.6 The Digital Signature Algorithm

The Digital Signature Algorithm, contained in the Digital Signature Standard [NIS94], has been derived from the ElGamal signature scheme and the Schnorr signature scheme. The public and secret keys are $y = g^x$ and $x \in_R \mathbb{Z}_q$, respectively. A signature $(R, s) \in \mathbb{Z}_q \times \mathbb{Z}_q$ for a document $M \in \{0, 1\}^*$ is generated by computing:

$$R = f(g^r), \quad s = r^{-1}(\mathcal{H}(M) + xR), \quad r \in_R \mathbb{Z}_q$$

where $\mathcal{H} : \{0, 1\}^* \to \mathbb{Z}_q$ is cryptographic hash function and f a deterministic one-way function as mentioned in Section 2.2.4. In the version presented in [NIS94], $\mathbb{G}_q$ is a subgroup of $\mathbb{Z}_p^*$, and f the reduction modulo q.

The verification of (R, s) with respect to M and y is done as follows:

$$R \stackrel{?}{=} f(g^{\mathcal{H}(M)s^{-1}} y^{Rs^{-1}})$$

2.2.7 The RSA Encryption and Signature Scheme

In [RSA78], the first public key encryption scheme has been proposed, which is secure under the assumption that factoring the product of two large primes is hard. During the key generation process, one computes $n = pq$, where p and q are distinct primes, chosen appropriately. Then, e is chosen at random from $\mathbb{Z}_{\varphi(n)}^*$, where φ is Euler's totient function. Furthermore, d is computed, such that the following holds:

$$ed \equiv 1 \pmod{\varphi(n)}$$

The public key is then (e, n) and (d, n) is the corresponding secret key. Normally p and q are discarded at the end of the key generation process. The encryption of a plaintext message $m \in \mathbb{Z}_n$ is performed by one modular exponentiation:

$$c = m^e \text{ MOD } n$$

Given the secret key (d, n) and c, the plaintext m can be obtained by calculating:

$$m = c^d \text{ MOD } n$$

In its original form, the RSA encryption scheme is deterministic and hence insecure for applications where the set of possible plaintexts is small (e.g. electronic voting, cf. Section 2.6). A probabilistic variant of RSA encryption, also known as RSA-OAEP, can be found in [BR95] and is described later on in Section 3.3.3.3.

RSA encryption can also be used as a signature scheme. Here, (d, n) is the signature key and (e, n) the corresponding verification key. $M \in \{0, 1\}^*$ is signed as follows:

$$s = \mathcal{H}(M)^d \text{ MOD } n$$

where $\mathcal{H} : \{0,1\}^* \rightarrow \mathbb{Z}_n$ is a cryptographic hash function. The verification of s with respect to M and (e, n) is performed straightforwardly as follows:

$$\mathcal{H}(M) \stackrel{?}{=} s^e \text{ MOD } n$$

It is important to sign the hash value to prevent existential forgery.

2.3   Universally Unique Identifiers

Uniquely identifying physical or virtual objects by means of a binary string, also known as identifier, is an intrinsic requirement for modern IT-systems. Obviously, it is important that within the life-time of the system, two different objects are not assigned the same identifier. Otherwise, the system cannot distinguish between these objects, which may lead to system failures or security holes.

As long as the objects remain in the proximity of the issuing party, there is no problem at all. Unique identifiers can easily be generated. A naive approach would be to use a counter which is incremented before issuing a new identifier. In a distributed environment, where several issuing parties generate identifiers for objects (that in turn will be exchanged with other instances of the system), the above mechanism must be enhanced: each issuing party will hold a system-wide unique identifier, which will be concatenated to the counter in order to derive several system-wide unique identifiers.

Unfortunately, uniqueness is often not the only requirement. In many applications privacy protection in terms of anonymity and unlinkability is required. In this context anonymity means that a malicious party cannot determine the issuer of a given identifier efficiently. Unlinkability implies that it cannot be determined if two given identifiers have been generated by the same issuer.

ITU-T Recommendation X.667 [IT04] and ISO/IEC 9834-8:2005 [ISO05], are the standards for Universally Unique Identifiers (UUIDs). They are based on the RFC 4122 [MS05], which addresses both uniqueness and privacy protection. UUIDs are implemented as Globally Unique Identifiers (GUIDs) by the Microsoft Corporation. In practice, UUIDs can be found in a large number of applications, including:

- Object identifiers in various programming languages such as the J2SE 5.0 release of Java [Java], C# [DAN02], PHP [PHP], .NET [Ric02], Javascript [Javb], ColdFusion [Col], Perl [Per], Tcl [Tlc], Python [Pyt] and Ruby [Rub],

- Identifiers in the Microsoft Windows Registry,

- Identifiers used in databases, such as MySQL [MyS],

- Identifiers used in XML [XML], and

- Identifiers used in RPCs (remote procedure calls) [ISO96].

All versions of UUIDs, generated according to the RFC 4122, are 128 bit numbers. In [IT04], the three algorithms are specified for the generation of UUIDs. Hereby, the following promise is stated (cf. page 7 in [IT04]):

> *"Three algorithms are specified for the generation of unique UUIDs, using different mechanisms to ensure uniqueness."*

From a technical point of view the term *ensure* suggests that no two runs of the used generation algorithm give the same output. This in turn means that no two UUIDs are identical. For the algorithms given in [IT04, MS05, ISO05] this is *totally wrong*:

Algorithm 1: Uniqueness is guaranteed for UUIDs version 1 if IEEE 802 MAC-addresses are used and none of them are manipulated. In the case that no network address is available, an address is chosen at random. Hence, uniqueness is not guaranteed unless only real network addresses are used. However, privacy is not preserved since the UUID then does not hide the MAC-address.

Algorithm 2: UUIDs are derived from unique names using cryptographic hash functions. MD5 [Riv05] is used for UUIDs version 3 and SHA-1 [NIS02] for UUIDs version 5. The birthday paradox suggests that a collision will occur with the probability of $1/2$ after $O(\sqrt{n})$ runs, where n is the largest possible output number [MVO96]. Privacy is preserved due to the one-way property of cryptographic hash functions.

Algorithm 3: UUIDs version 4 are generated at (pseudo-)random, i.e. collisions can again occur. Privacy is preserved since no name or network address is involved.

It is obvious that none of the proposed algorithms ensures uniqueness while *at the same time* preserving the privacy of the issuer. Notice that in the literature, it is often mentioned that UUIDs are not truly unique, but rather than a collision is accepted to occur with a low probability. Whether complete uniqueness is intrinsic certainly depends on the application for which the UUIDs are used.

2.4 Pseudonym Systems

Using basic authentication schemes one instance can prove its identity to another instance. Advanced applications, however, require that users remain anonymous as long as they behave honestly. In such cases pseudonym systems are useful [Cha81]. There are several common types of pseudonyms, such as person pseudonyms, role-relationship pseudonyms and transaction pseudonyms [PK01]. The more pseudonyms are used (e.g. transaction pseudonyms), the lower the linkability of different sessions becomes assuming pseudonyms are chosen independently. An obvious requirement is that pseudonyms are *globally unique* [Cha81, PK01].

In principle, a pseudonym is a unique identifier where the linking to the corresponding holder is hidden. In several applications, it is necessary that in case of emergency the linking can be disclosed by a trusted party. If a pseudonym is not unique, the holder might not be uniquely identifiable. Hence, the same problems arise as were discussed in Section 2.3. This affects generation procedures where pseudonyms are generated at random and no mechanisms are employed to avoid collisions. Vulnerable applications include [CE87, KRJ98] and [SS01], for instance. Other applications do not specify how pseudonyms are generated. In [CL01], pseudonyms are of the form $N_1||N_2$, where N_1 is chosen by the user and N_2 by an organization. No details are given on how (e.g. at random) the two parts N_1 and N_2 are chosen. To ensure the unique identification of a user, the organization might be responsible for ensuring that N_2 is unique.

2.5   Key Generation Algorithms

Key generation algorithms are based on random or pseudo-random number generators and hence collisions can naturally occur. A collision can be dangerous if it is publicly detectable. For most private key cryptosystems this is not the case. In public key cryptosystems collisions among public keys are publicly detectable. In this section we show that one might take advantage of this fact.

2.5.1   Private Key Cryptosystems

To provide confidentiality and integrity, symmetric cryptosystems are widely used, particularly since they are also very efficient. In such systems, the sender and the receiver have to agree on some secret information which must not be known by any other (unauthorized) instance. Now consider the following situation: two secret keys k and k' have been generated through a key agreement protocol. For example, Alice and Bob hold k and Carol and Don hold k'. If $k = k'$, then Carol and Don can perform actions on behalf of Alice and Bob, and vice versa. For typical applications, such as confidential communication through the Internet, such a collision is not practically detectable and seems to cause no vulnerabilities. But there exist particular applications where such a collision is problematical, even though a symmetric scheme is used: remote controls for car doors. Imagine the situation where a person walks to his car, activates his remote control and the car besides his own one opens the door. This is publicly detectable, because the reaction of the car is noticeable. Before remote controls became popular, one would have to go to a particular car in order to find out if his key could have been used to open it. With a remote control this can be done without touching the car.

2.5.2 ElGamal-like Encryption and Signature Generation

In ELGamal-like schemes, secret numbers are hidden using one modular exponentiation, e.g. for a secret key z the corresponding public key is $h = g^z$. This happens to be the case in the original ElGamal encryption and signature scheme, the Schnorr signature algorithm, the Digital Signature Algorithm and the simplified Cramer-Shoup encryption scheme. Given two public keys h and h', one can follow from $h = h'$, that $z = z'$, because group exponentiation is injective, and $z, z' < q$. This obviously leads to security breaches, since the holders of h and h' can mutually impersonate each other.

2.5.3 Cramer-Shoup Encryption

In the Cramer-Shoup encryption scheme, the public key components are $c = g_1^{x_1} g_2^{x_2}$, $d = g_1^{y_1} g_2^{y_2}$ and $h = g_1^{z_1} g_2^{z_2}$. Let $pk = (c, d, h)$ and $pk' = (c', d', h')$ be the public keys of Alice and Bob. Assume that Bob has access to pk and any ciphertext (u_1, u_2, e, v) that has been generated by Alice. The following can be shown:

1. *Violation of the Non-Malleability:* If any $\delta \in \{c', d', h'\}$ is equal to c or d, then Bob can modify (u_1, u_2, e, v) efficiently to a different *valid* ciphertext, whose corresponding plaintext is meaningfully related to m.

2. *Violation of the One-Way Property:* If any $\delta \in \{c', d', h'\}$ is equal to h, then Bob is able to decrypt (u_1, u_2, e, v) efficiently.

In the ElGamal encryption scheme, the correct decryption of a ciphertext is guaranteed by the fact that if $h = g^z$ and $u = g^r$, then $h^r = u^z$, i.e. the commutativity of group exponentiation. The correctness of the Cramer-Shoup scheme relies on the same principle. However, here the public key component for the encryption is $h = g_1^{z_1} g_2^{z_2}$, where $dlog_{g_1}(g_2)$ is unknown. To provide a correct decryption, both $u_1 = g_1^r$ and $u_2 = g_2^r$ are necessary, such that $h^r = u_1^{z_1} u_2^{z_2}$ holds. For simplicity, we state the following, to which we refer to when showing a weakness.

Lemma 2.5.1 *Let $a = g_1^{a_1} g_2^{a_2}$, $b = g_1^{b_1} g_2^{b_2}$, $u_1 = g_1^r$, $u_2 = g_2^r$, where $a_1, a_2, b_1, b_2, r \in \mathbb{Z}_q$. If $a = b$ then $u_1^{a_1} u_2^{a_2} = u_1^{b_1} u_2^{b_2}$.*

Proof. Since $a = b$ it is obvious that $a^r = b^r$. Now set $a = g_1^{a_1} g_2^{a_2}$ and $b = g_1^{b_1} g_2^{b_2}$. Then $a^r = (g_1^{a_1} g_2^{a_2})^r = g_1^{a_1 r} g_2^{a_2 r} = u_1^{a_1} u_2^{a_2}$ and analogously $b^r = u_1^{b_1} u_2^{b_2}$. $\square$

2.5.3.1 Violation of the Non-Malleability

Before decrypting (u_1, u_2, e, v), Alice has to verify if $v = u_1^{x_1} u_2^{x_2} (u_1^{y_1} u_2^{y_2})^{\alpha}$ holds, where $\alpha = \mathcal{H}(u_1, u_2, e)$. From the ElGamal encryption scheme we know that e can be simply modified to e', such that the decryption of e' results in a plaintext m' meaningfully related to m. In the Cramer-Shoup scheme, this manipulation is in principle also possible, but is detectable before decryption takes place. To cope with this problem, the attacker Bob hence needs to modify the verification basis v as well, such that the manipulation remains undetected.

If at least one $\delta \in \{c', d', h'\}$ exists, such that $\delta \in \{c, d\}$, then the non-malleability of the Cramer-Shoup encryption scheme can be violated as follows.

Let $\delta = g_1^{\delta_1} g_2^{\delta_2}$, where $\delta_1, \delta_2 \in \mathbb{Z}_q$, so that Bob can efficiently detect if any $\delta \in \{c', d', h'\}$ is equal to c or d. Through sk' he knows δ_1 and δ_2 and computes $e' = eb$ and $\alpha' = \mathcal{H}(u_1, u_2, e')$. To obtain a valid verification basis v', he has two options:

1. If $\delta = c$, then he computes $v' = u_1^{\delta_1} u_2^{\delta_2} (v (u_1^{\delta_1} u_2^{\delta_2})^{-1})^{\alpha' \alpha^{-1}}$. From Lemma 2.5.1 it follows, that $u_1^{x_1} u_2^{x_2} = u_1^{\delta_1} u_2^{\delta_2}$. So, the validity of v' holds:

$$v' = u_1^{\delta_1} u_2^{\delta_2} (v(u_1^{\delta_1} u_2^{\delta_2})^{-1})^{\alpha' \alpha^{-1}} = u_1^{x_1} u_2^{x_2} (c^r d^{r\alpha} (u_1^{x_1} u_2^{x_2})^{-1})^{\alpha' \alpha^{-1}}$$
$$= c^r (c^r d^{r\alpha} (c^r)^{-1})^{\alpha' \alpha^{-1}} = c^r (d^{r\alpha})^{\alpha' \alpha^{-1}} = c^r d^{r\alpha \alpha' \alpha^{-1}} = c^r d^{r\alpha'}$$

2. If $\delta = d$, then he computes $v' = v(u_1^{\delta_1} u_2^{\delta_2})^{\alpha' - \alpha}$. From Lemma 2.5.1 it follows, that $u_1^{y_1} u_2^{y_2} = u_1^{\delta_1} u_2^{\delta_2}$. So, the validity of v' holds:

$$v' = v(u_1^{\delta_1} u_2^{\delta_2})^{\alpha' - \alpha} = c^r d^{r\alpha} (u_1^{y_1} u_2^{y_2})^{\alpha' - \alpha} = c^r d^{r\alpha} d^{r(\alpha' - \alpha)} = c^r d^{r\alpha + r\alpha' - r\alpha} = c^r d^{r\alpha'}$$

If $\delta, \delta' \in \{c', d', h'\}$, such that $\delta = c$ and $\delta' = d$, then v' can be computed without v:

$$v' = u_1^{\delta_1} u_2^{\delta_2} (u_1^{\delta_1'} u_2^{\delta_2'})^{\alpha'} = c^r d^{r\alpha'}$$

The new valid ciphertext is (u_1, u_2, e', v'). During decryption, the verification of v' succeeds for obvious reasons. Thus, the decryptor computes $e'(u_1^{z_1} u_2^{z_2})^{-1}$ and gets $m' = mb$, which is a plaintext meaningfully related to m.

We conclude that no matter which of the public key components of pk' collides with c or d, violation of the non-malleability property results.

2.5.3.2 Violation of the One-Way Property

If the verification of the ciphertext (u_1, u_2, e, v) succeeds, Alice can obtain m by computing $e(u_1^{z_1} u_2^{z_2})^{-1}$. Bob, who wants to get m, needs to remove $u_1^{z_1} u_2^{z_2}$ from e.

If at least one $\delta \in \{c', d', h'\}$ exists, such that $\delta = h$, then sk' can be used to successfully obtain m. This can be shown as follows.

Let $\delta = g_1^{\delta_1} g_2^{\delta_2}$, where $\delta_1, \delta_2 \in \mathbb{Z}_q$. Bob can then efficiently detect if any $\delta \in \{c', d', h'\}$ is equal to h. Through sk' he knows δ_1 and δ_2. From Lemma 2.5.1 it follows that he can obtain m by computing $e(u_1^{\delta_1} u_2^{\delta_2})^{-1}$. The correctness holds since:

$$e(u_1^{\delta_1} u_2^{\delta_2})^{-1} = mh^r(u_1^{z_1} u_2^{z_2})^{-1} = mh^r(h^r)^{-1} = m$$

Thus, if any public key component of pk' is equal to h, then Bob can get m by running the standard decryption algorithm. Hence, in contrast to breaking the non-malleability, Bob does not need to perform any special computations to obtain m.

To overcome the above attacks, one has to ensure that the public key components are system-wide unique. In the standard Cramer-Shoup scheme, this is only possible if $dlog_{g_1}(g_2)$ is known by the key generation algorithm. In the simplified variant, it suffices to ensure that the secret key components are system-wide unique.

2.5.4 RSA-like Cryptosystems

As stated by the fundamental theorem of arithmetic, every natural number (except 1) can be written as a product of prime-powers, which is a unique representation. The RSA cryptosystem [RSA78], the Rabin cryptosystem [Rab79], the Pailler cryptosystem [Pai99] and the Cramer-Shoup signature scheme [CS99] are based on the problem of factoring large prime-products, for instance. An intrinsic requirement for these schemes is that the following is prevented for two moduli $n = pq$ and $n' = p'q'$, where p, q, p' and q' are primes, of different users [Sim83, DeL84]:

1. *Common Prime Attack:* If p or q equals to p' or q', the holder of n' can factor n efficiently, since $gcd(n, n') \in \{p', q'\}$.

2. *Common Modulus Attack:* If $n = n'$, then $p = p'$ and $q = q'$ or $p = q'$ and $q = p'$. Notice that having a public and the corresponding secret key with respect to n one knows a multiple of $\varphi(n)$. This might lead to a successful factorization of n [Sal90].

The above problems also exist if special RSA-moduli for fast computations are used [CHLS97, Tak98]. Thus, one can see that it is important that the prime-factors of all moduli that are used in the system are system-wide unique.

2.6 ElGamal-like Encryption

During the encryption of a plaintext m, a randomizer r is used to provide the indistin-guishability of ciphertexts [GM82, GM84]. To make the encryption process invertible by use of the secret key z, $u = g^r$ is included in the ciphertext. Now consider two ciphertexts $(u, e) = (g^r, mh^r)$ and $(u', e') = (g^{r'}, m'h^{r'})$, that have been generated by use of the same public key h. If $u = u'$, then $r = r'$ and hence the equality of r and r' is publicly detectable. Computing $e'e^{-1}$ gives

$$e'e^{-1} = m'h^{r'}(mh^r)^{-1} = m'h^r m^{-1}h^{-r} = m'm^{-1}$$

Whether m and m' can be extracted from $m'm^{-1}$ depends on how m and m' are encoded. If they are random numbers, it is probably impossible to find m' and m. However, if the set of valid plaintexts is very small, then m' and m might be efficiently found. A typical example, where only a few plaintexts are involved, is an electronic voting scheme. In [CS97], a voting scheme was proposed where ballots are encrypted by the use of a modified ElGamal encryption. The scheme requires a special homomorphic property so that votes can be added in encrypted form. This leads to the convenient fact that only one ciphertext (containing the result) needs to be decrypted after the tally. If the space of plaintexts is very small, the ElGamal scheme can be modified such that it provides additive homomorphic properties whilst remaining feasible to decrypt. Let γ be a generator of $\mathbb{G}_q$, such that $dlog_g(\gamma)$ is unknown. Moreover, let $M \subset [0, q-1]$ be the set of possible plaintexts, with the condition that for a sum s of k plaintexts, computing s from γ^s is feasible. In this setting, the ElGamal ciphertext corresponding to the plaintext $m \in M$ is defined as follows:

$$(u, e) = (g^r, \gamma^m h^r), \quad r \in_R \mathbb{Z}_q$$

Now, given two ciphertexts $(u, e) = (g^r, \gamma^m h^r)$ and $(u', e') = (g^{r'}, \gamma^{m'} h^{r'})$, where $m, m' \in M$, one can compute the ciphertext (u'', e'') of the plaintext $m + m'$:

$$u'' = uu' = g^r g^{r'} = g^{r+r'}, \quad e'' = ee' = \gamma^m \gamma^{m'} h^r h^{r'} = \gamma^{m+m'} h^{r+r'}$$

The decryption of (u'', e'') results in $\gamma^{m+m'}$. If M is a small set, then $dlog_\gamma(\gamma^{m+m'})$ can be efficiently and uniquely found by brute force. In the context of electronic elections, M is the set of possible votes. If properly encoded, the sum of millions of votes is still small enough that computing or finding the discrete logarithm is easy.

Now consider the voting scheme of [CS97], where h is the public key of the election authority. For simplicity reasons, consider an election concerning two parties, where votes are binary, i.e. a vote v is chosen from the space $M = \{1, -1\}$. An election with respect to two parties A and B can take place under the following assumptions:

1. $v = 1$ means 'yes' for party A and 'no' for party B, whereas

2. $v = -1$ means 'no' for party A and 'yes' for party B.

A voter encrypts $v \in \{1, -1\}$ as follows, resulting in the encrypted ballot (u, e):

$$(u, e) = (g^r, \gamma^v h^r), \quad r \in_R \mathbb{Z}_q$$

To provide robustness, each voter has to prove that $v \in \{1, -1\}$, without revealing v. Therefore, a Σ-OR-proof for discrete logarithms is necessary. The corresponding protocol can be found in [CS97]. A similar protocol is discussed in Section 5.2.3.3.

A requirement for voting schemes is that the vote of a particular voter cannot be efficiently obtained by one single party. In the above described voting scheme, this requirement is fulfilled by use of ElGamal threshold decryption [DF90]. The decryption key z is shared among a set of election authorities, so that up to a particular number of authorities can be corrupt without being able to decrypt a single vote. Honest authorities only agree to perform a decryption *after* having computed the ciphertext of the sum of all valid votes. Thus, the anonymity of the voter is effectively preserved.

Let v and v' be the votes of two different voters, where the same randomizer r has been used for encryption. Then we have the two ciphertexts (u, e) and (u', e'), where $u = u'$. Computing $e'' = e'e^{-1}$ gives the following possible results:

1. If $e'' = 1$, then $v = v' = 1$ or $v = v' = -1$.

2. If $e'' = \gamma^{-2}$, then $v' = -1$ and $v = 1$.

3. If $e'' = \gamma^2$, then $v' = 1$ and $v = -1$.

In the first case an attacker knows that both voters made the same decision, but not which one. In case 2 or 3, however, the exact decisions of the two voters can be identified. This contradicts the fundamental requirement that a vote cannot be linked to a voter in plain form.

In the Cramer-Shoup scheme, the encryption of plaintext works similarly to that of the standard ElGamal encryption scheme. Hence, it suffers from the same drawbacks as shown above. Let (u_1, u_2, e, v) and (u_1', u_2', e', v') be two ciphertexts, computed by use of the same public key component h. If $u_1 = u_1'$, then $r = r'$ and thus computing $e'e^{-1}$ results in $m'm^{-1}$.

If one chooses r to be system-wide unique, then the above attack is no longer relevant.

2.7   ElGamal-like Signature Generation

In the ElGamal encryption scheme, the use of the same randomizer several times, together with the same public key, possibly leads to the disclosure of some information about the plaintext. In the ElGamal signature scheme, using the same randomizer more than once with the same secret key leads to the full disclosure of the secret key. Let $(R, s) = (g^r, r^{-1}(\mathcal{H}(M) - xf(R)))$ and $(R', s') = (g^{r'}, r'^{-1}(\mathcal{H}(M') - xf(R')))$ be two ElGamal signatures of two distinct hash values $\mathcal{H}(M)$ and $\mathcal{H}(M')$, generated by use of the same secret key x. If $R = R'$, then $r = r'$ holds and the following can be computed (we set $m := \mathcal{H}(M)$ and $m' := \mathcal{H}(M')$ for simplicity):

$$
\begin{aligned}
sm' - s'm &= r^{-1}(m - xf(R))m' - r^{-1}(m' - xf(R))m \\
&= r^{-1}mm' - r^{-1}xf(R)m' - r^{-1}m'm + r^{-1}xf(R)m \\
&= r^{-1}xf(R)(m - m') := A \\
s - s' &= r^{-1}(m - xf(R)) - r^{-1}(m' - xf(R)) \\
&= r^{-1}(m - xf(R) - m' + xf(R)) \\
&= r^{-1}(m - m') := B
\end{aligned}
$$

If $f(R) \neq 0$, then $A \neq 0$ and $B \neq 0$, since r has an inverse in $\mathbb{Z}_q^*$ and $m \neq m'$ holds by assumption. Now, computing $A(Bf(R))^{-1}$ results in x. A similar attack can be mounted if two different users use the same two different randomizers for signature generation. Then, one can solve a linear system of four equations with four unknowns to obtain the secret key. If one chooses r to be system-wide unique, the above attacks are not relevant anymore.

2.8 Interactive Proofs of Knowledge

In an interactive proof of knowledge [BG93], an instance called the *prover* interactively proves the knowledge of a secret to an instance called the *verifier*. Thereby, he does not reveal any information about the secret. For instance, given an ElGamal public key h and the generator g, the prover convinces the verifier that he knows $dlog_g(h)$. To ensure that the proof works correctly, the verifier must have authentic access to some public information that (uniquely) defines the secret knowledge. Besides other properties (cf. [BG93] and Section 5.2.3), an interactive protocol is a proof of knowledge (or in a special case Σ-proof [Cra96]) if a so-called knowledge extractor can be given. For Σ-proofs, such a knowledge extractor informally works as follows: if a prover sends the same first message t (computed from the same randomness) twice and then correctly responds to two different challenges c and c' by sending s and s', then the secret information must be efficiently extractable from the two protocol triples (t, c, s) and (t, c', s'). Obviously, in real protocol runs such a situation must not happen or should at least be undetectable, since otherwise a dishonest verifier can extract the secret. In the following, we sketch the Σ-proof for the knowledge of a discrete logarithm [Sch89]:

1. The prover sends $t = g^\alpha$, where $\alpha \in_R \mathbb{Z}_q$, to the verifier.

2. The verifier returns a random challenge $c \in \mathbb{Z}_q$.

3. The prover responds with $s = \alpha - cx$.

4. The verifier checks if $t = y^c g^s$ holds.

As widely known, this protocol is a proof for the knowledge of $dlog_g(y)$. The corresponding knowledge extractor is the following: given the transcripts (t, c, s) and (t', c', s') of two rounds, where $t = t'$ and $c \neq c'$, x can be extracted as follows:

$$(s - s')(c' - c)^{-1} = ((\alpha - cx) - (\alpha - c'x))(c' - c)^{-1} = x(c' - c)(c' - c)^{-1} = x$$

Hence, if the prover selects $\alpha = \alpha'$ accidentally in any two runs of the above stated protocol (even with different verifiers), then the transcripts can be used to extract x. This problem can be overcome if α is generated to be system-wide unique. A knowledge extractor can still be given, but its appearance in practice is then avoided.

From the above it follows that the Schnorr signature scheme suffers from the same vulnerability. Notice that most interactive proofs of knowledge can be turned into digital signature schemes using the Fiat-Shamir paradigm (cf. Section 5.2.3.1).

COLLISION-FREE NUMBER GENERATION

Chapter 3

3.1 Introduction

In this chapter, several techniques are presented that enable the efficient local generation of numbers that are system-wide unique within a certain time-interval. The discussion remains at an abstract level allowing a wide range of applications to be covered. Furthermore, the algorithms presented satisfy additional requirements.

3.1.1 Integrating Uniqueness and Randomness

In the following we analyze what happens if the output of a random or pseudo-random number generator is required to be system-wide unique. Firstly, consider the following two abstract definitions [Yao82b, GGM86, MVO96]:

Definition 3.1.1 A *Random Number Generator* (RNG) generates numbers in a non-deterministic way, where each bit is generated independently and unbiased.

We refer to the output of a random number generator as (truly) random. In this context unbiased means that for each bit b of a block $\Pr[b = 0] = \Pr[b = 1] = 1/2$.

Definition 3.1.2 A *Pseudo-Random Number Generator* (PRNG) generates numbers in a deterministic way which appear to be random.

In the context of cryptography, the term *appear* tries to express the fact that by a polynomially bounded (or poly-bounded) algorithm, such numbers are not efficiently distinguishable from a truly random bit-sequence. To call a pseudo-random number generator cryptographically strong, the underlying pseudo-random bit generator needs to pass the next-bit test [Yao82b]: given an l-bit sequence, a poly-bounded algorithm cannot predict the $(l + 1)$-st bit with a probability greater than $1/2$.

If mechanisms are used within random number generators to avoid collisions, a generated number cannot be truly random anymore, since true randomness implies that bits are chosen independently and unbiased. This is clearly not the case because numbers that have already been generated may not be generated again. Nonetheless, the generation process can still be non-deterministic. For instance, consider a number generator having access to an oracle which is able to efficiently decide if a particular number has already been generated somewhere in the system. The number generator chooses a random number and queries the oracle which makes its decision and returns *accept* or *reject*. Unless each state of the whole infrastructure is stored, the generation process is not reproducible and hence non-deterministic, but the output cannot be truly random since some generated numbers are rejected by the oracle.

If numbers are generated in a non-deterministic way but appear to be random then the corresponding number generator has the first property of random number generators (i.e. non-deterministic) and the second property of pseudo-random number generators (i.e. appear to be random). It is neither a random number generator nor a pseudo-random number generator, but apparently something in-between. In [Sch01], such generators are referred to as *quasi-random number generators* to avoid confusion with other definitions within the context. Notice, however, that the properties *non-deterministic* and *appear to be random* do not imply uniqueness. Thus, in contrast to [Sch01], we state the following more general definition:

Definition 3.1.3 A *Quasi-Random Number Generator* (QRNG) generates numbers in a non-deterministic way which appear to be random.

Remark 3.1.1 The term *quasi-random* is also used with a totally different meaning in the context of Monte Carlo methods [Nie92].

Table 3.1 summarizes the basic properties of the defined number generators. Notice that truly random number generators cannot exhibit deterministic behavior. This is impossible to achieve since a determinism implies dependency between the numbers.

	non-deterministic	deterministic
truly random	RNG	contradiction
appears to be random	QRNG	PRNG

Tab. 3.1: Properties of Number Generators.

It is clear that a (quasi-)random number generator, being provided with mechanisms to avoid collisions, is at most a quasi-random number generator, but never a random number generator. It remains to be analyzed what happens if uniqueness is required from a pseudo-random number generator. As mentioned, including a requirement of uniqueness results in numbers which at most appear to be random, but the generation process can still be deterministic, so that one gets a pseudo-random number generator.

Definition 3.1.4 Let Δ be a time-interval. The output o of a generator is called

1. *locally unique*, if within Δ, for every output o' of the same generator $o \neq o'$ holds.

2. *globally unique*, if within Δ, for every output o' of any other generator $o \neq o'$ holds.

3. *system-wide unique*, if within Δ, o is locally and globally unique.

In general, we say o is *unique*, if one of the above three definitions holds.

The constructions given later on guarantee that a generated sequence of numbers is system-wide unique. Interestingly, if the time-interval is extended beyond Δ, the numbers are no longer system-wide unique, but at least globally unique.

Definition 3.1.5 A *Collision-Free Number Generator* (CFNG) generates numbers which are unique and appear to be random.

This is a very abstract definition of a collision-free number generator, since it also covers solutions which ensure local uniqueness. In the remainder of this chapter, we focus on the most difficult case, where the output is required to be system-wide unique.

3.1.2 Requirements

For several applications, such as the ones presented in Chapter 2, uniqueness is not the only requirement for a generated number. Given a Universally Unique Identifier, for instance, one must not be able to efficiently identify the corresponding issuer. Hence, there is the additional requirement for privacy. In the context of key generation, a generated number must provide a certain quality of randomness. To be of practical relevance, the generation process must also be efficient. Thus, the requirements are:

R1 (Uniqueness): A generated number is system-wide unique.

R2 (Randomness): A generated number appears to be random.

R3 (Efficiency): The generator is efficient regarding communication, time and space.

R4 (Privacy): Here the following three cases are distinguished:

(a) *Hiding*: Given a generated number, a poly-bounded algorithm is not able to efficiently identify the corresponding generator.
(b) *Unlinkability*: Given a set of generated numbers, a poly-bounded algorithm is not able to efficiently decide which of them come from the same generator.
(c) *Independency*: Breaking the hiding-property of one particular generator does not lead to an efficient breaking of (a) or (b) of any other generator of the system.

The main goal of this work is to provably fulfill R1. For R2, it is sufficient that a sequence of generated numbers is computationally indistinguishable from a true random sequence. If not, the quality of randomness has to be verified empirically through statistical evaluations with respect to the application for which the generator is used. R3 is threefold: the complexity of a generation process is normally measured through its computational costs and the space used. In the case of collision-free number generation, communication also needs to be taken into account. Otherwise, a trusted party could be used to decide if a number has already been generated somewhere in the system, which is at most of theoretic interest. R4 can be seen as an *optional requirement* which is especially important for privacy-preserving applications. If transaction pseudonyms are generated through a collision-free number generator, for instance, then it must be obviously infeasible to identify the corresponding issuer. This property we refer to as hiding. A stronger assumption is to require that numbers, which have been generated by the same generator, are not mutually linkable. This property we refer to as unlink-ability. As a final requirement the breaking of the hiding- or unlinkability-property of one generator must not lead to the breaking of the hiding- or unlinkability-property of all other generators. This requirement for independency excludes all approaches using common master secrets. Using master secrets is very dangerous if no tamper resistant devices are involved since privacy can only be maintained against passive adversaries. Such adversaries do not deviate from the specification.

Remark 3.1.2 Fulfilling (a) and (b) of R4 generally requires the reduction to a mathematical problem. If such a reduction can be given, then R2 may be provably fulfilled as a by-product. One might think of a Diffie-Hellman Triple, for instance, which is indistinguishable from a random triple under the Decision Diffie-Hellman Assumption [Bon98]. Hence, requirements R2 and R4 seem to be closely related.

3.1.3   Related Work

Despite being an interesting research topic, only a few results can be found in litera-ture. These are compared with our results in Table 3.2 with respect to the requirements stated in the previous section. The last three rows represent our contributions. For R3, we distinguish between R3.C (communication), R3.T (time) and R3.S (space), and for R4 we distinguish between R4.H (hiding), R4.U (unlinkability) and R4.I (in-dependency). In the following, each approach is sketched on an abstract level.

Universally Unique Identifiers: As far as unique number generation is concerned, the use of Universally Unique Identifiers [MS05] seems to make sense. As discussed in Section 2.3, none of the three standardized algorithms (cf. 1, 2, 3a and 3b below) ensures uniqueness, while at the same time protecting the privacy of the issuer:

$$1: \quad o = \textit{MAC-address}||\textit{current time} \qquad 3\text{a}: \quad o = \mathsf{RNG}()$$
$$2: \quad o = \mathcal{H}(\textit{unique name}) \qquad\qquad\quad\ 3\text{b}: \quad o = \mathsf{PRNG}()$$

The first algorithm just concatenates the MAC-address of the computer to the current

	Requirement							
Reference	R1	R2	R3.C	R3.T	R3.S	R4.H	R4.U	R4.I
[MS05].1	yes	none	once	little	little	none	none	high
[MS05].2	no	high	once	little	little	high	high	high
[MS05].3a	no	perfect	once	medium	little	perfect	perfect	perfect
[MS05].3b	no	high	once	little	little	high	high	high
[IPC02]	yes	none	often	little	little	–	–	high
[Hor98]	yes	high	once	little	little	high	high	none
[HS98]	yes	high	once	little	little	high	high	none
[HSW98a]	yes	high	once	little	little	high	high	none
[HSW98b]	yes	high	once	little	little	high	high	none
[Sch01]	yes	high	once	little	little	none	none	high
[SS05]	yes	high	once	much	much	high	high	high
[SSR07b].1	yes	high	once	medium	medium	high	high	high
[SSR07b].2	yes	high	once	medium	medium	high	high	high
[SSR07b].3	yes	high	once	medium	little	high	high	high

Tab. 3.2: Related Work in Comparison with our Contribution for Public Outputs.

time. Assuming that MAC-addresses are ideal (i.e. not cloned or manipulated), each output is ensured to be unique. However, the issuer can be identified since the MAC-address is accessible. Moreover, no randomness is achieved. The second algorithm computes a cryptographic hash value of a unique name. This approach is very efficient and provides a high degree of privacy, but uniqueness is not guaranteed. If the third algorithm is used we distinguish two cases: if a random number generator is used, then the algorithm performs poorly with respect to time, but a high degree of privacy and randomness is provided. On the other hand, if a pseudo-random number generator is used, then the performance can be substantially better, but the degree of randomness and privacy is slightly lower. In both cases, uniqueness is not guaranteed.

Global Positioning System: A different approach can be found in [IPC02]. The system concatenates the current time to the physical position of a device, which has been obtained by use of the Global Positioning System (GPS), creating an output

$$o = physical\ position \| current\ time$$

that is guaranteed to be unique. This approach obviously achieves no randomness in the output. Moreover, GPS is not very precise and is not always available (e.g. within buildings), leading to a low fulfillment of R3.C. Privacy-protection is also questionable.

If several unique numbers are generated within a short period of time then the numbers are linkable if the physical position has remained constant. If the GPS is used together with a mobile phone, anonymity might not be guaranteed anymore.

Symmetric Encryption with a Master Secret: One solution, which overcomes the mentioned problems, is the local generation of system-wide unique private keys, as proposed in [Hor98] and modified in [HS98, HSW98a] and [HSW98b]. Each generator encrypts a unique number u (e.g. unique identifier concatenated to a counter) with a master secret MS using a symmetric encryption function E_S:

$$o = E_S(u, MS)$$

This approach provides uniqueness and a high degree of randomness, but violates the requirement for independency. Once the master secret has been found, all keys generated by the system (or with respect to this master secret) can be linked to their originators. Moreover, the master secret needs to be protected (e.g. stored in a tamper-resistant device), otherwise hiding and unlinkability cannot be guaranteed because a malicious holder of a generator can misuse the master secret.

Symmetric Encryption without a Master Secret: The approach in [Sch01] is an improvement of the approach in [Hor98]. The unique number u is encrypted by a locally chosen random key k, which is then concatenated to the output:

$$o = E_S(u, k) \| k$$

Independency is achieved and randomness is still largely preserved. Moreover, uniqueness is guaranteed (we prove this in Section 3.2.3). However, since a symmetric encryption scheme is used one can easily invert the generation process and obtain u. Hence, hiding and unlinkability can only be achieved if o is kept secret.

Asymmetric Encryption without a Master Secret: The solution in [SS05] is very similar to the one proposed in [Sch01]. The only difference is, that an asymmetric encryption function E_A is used, where the public key pk is chosen locally at random (and according to possibly existing conditions) and finally concatenated to the output:

$$o = E_A(u, pk) \| pk$$

Uniqueness is achieved using the same idea as in [Sch01], but the generation process can no longer be efficiently inverted. Thus, privacy is secured even if o is used in public. Unfortunately, the output of the generator is quite large and the generation process is less efficient than if symmetric encryption is used.

Our Approaches: In this chapter, the approaches presented in [Sch01] and [SS05] are generalized and extended with respect to the requirements stated in Section 3.1.2.

This gives three solutions for collision-free number generation. Notice that a preliminary abstract of this chapter has been published in [SSR07b]. Hence, in Table 3.2, our three approaches are referred to as [SSR07b].1, [SSR07b].2 and [SSR07b].3. Our constructions differ in some properties, but provide provable uniqueness and sufficiently fulfill the other requirements if applied appropriately.

Notice that privacy is optional in the case where the output is *publicly* accessible. This, however, is not the case if the generated number is a secret key, for instance.

3.1.4 Contribution and Organization

As mentioned, the contribution of this work is the generalization of the approaches given in [Sch01] and [SS05]. Moreover, this work is the first to formulate requirements such that a large range of applications can be covered. In particular, the requirement for privacy allows the use of the generation concepts for digital pseudonyms and Universally Unique Identifiers. Asides from giving an abstract design, the most efficient implementation is proposed with respect to the length of a generated number, whilst still fulfilling all other requirements.

The remainder of this chapter is organized as follows: in Section 3.2, the core principle is generalized. Then, in Section 3.3, selected techniques to provide flexibility for the design of a generator are discussed. In Section 3.4, three types of collision-free number generators are defined, which are then instanced by some practical functions in Section 3.5. Finally, the applications that were given in Chapter 2 are revisited.

3.2 Basic Construction

In this section, a basic method is given for locally generating numbers, such that they are system-wide unique. This basic solution *ensures* uniqueness while preserving a certain level of randomness if the used building blocks are chosen carefully.

3.2.1 Involving Randomness

First of all, a source is required which outputs random, quasi-random or pseudo-random numbers of a certain length. For the sake of generality, the quality of randomness is not specified here. Thus, such a source is simply called number generator:

Definition 3.2.1 Let $R = \{0,1\}^l$. The *Number Generator* NG is defined as follows:

$$NG : \emptyset \rightarrow R, \quad NG() = r$$

where r is chosen at random, quasi-random or pseudo-random.

A collision-free number generator uses a number generator, such as the one defined above, as a building block. The abstractness of the definition leads to a better coverage of the possible variants of a collision-free number generator (cf. Table 3.3).

	non-deterministic	deterministic
truly random	contradiction	contradiction
appears to be random	QRNG	PRNG

Tab. 3.3: Properties of Collision-Free Number Generators.

Keep in mind that by using a random or quasi-random number generator for NG, one obtains a *Collision-Free Quasi-Random Number Generator*, which is quite useful for non-reproducible cryptographic purposes (e.g. signature keys). Alternatively, taking NG as a pseudo-random number generator leads to the implementation of a *Collision-Free Pseudo-Random Number Generator*. Such a solution might be interesting for applications where one needs synchronicity of processes (e.g. remote controls for cars).

3.2.2 Uniqueness Generation

Apart from the number generator, a function is needed, which outputs a system-wide unique number every time it is run. Here, uniqueness is the only property required from the output, i.e. there is no assumption concerning its probability distribution.

Definition 3.2.2 The *Uniqueness Generator* UG is defined as follows:

$$\text{UG} : \emptyset \rightarrow U, \quad \text{UG}() = u$$

where u is system-wide unique.

UG can be instanced in several ways. An efficient solution is to initialize each uniqueness generator with a system-wide unique identifier UI. Then, during every run of UG, a system-wide unique number is derived from UI. Since no requirements are set for the probability distribution of the output, this can be achieved by incrementing a counter which is concatenated to the identifier. More details of such a simple approach are given in Section 3.3.2. Notice that this way to generate u is very efficient, because no communication is necessary after the initialization process.

3.2.3 Uniqueness Randomization

A first step to provide randomness is to transform u, such that the resulting block appears to be random, whilst still preserving the uniqueness-property. The idea is that u is randomized by a randomizer r such that the output is (computationally) indistinguishable from a random string, or at least provides sufficient randomness for the particular application. Practical candidates that we recommend include mixing-transformations according to Shannon [Sha49] (e.g. symmetric encryption) and injective one-way functions that are believed to be hard to invert with respect to an unsolved mathematical problem (e.g. asymmetric encryption).

Definition 3.2.3 The *Uniqueness Randomization Function* f is defined as follows:

$$f : U \times R \to O_f, \quad f(u,r) = o_f$$

where f is injective in u for all $r \in R$.

f_r denotes that the second argument of f is an arbitrary but fixed $r \in R$. Unfortunately, the output of f is not guaranteed to be unique. Let $u, u' \in U$, $u \neq u'$ and $r, r' \in R$. Then the following two cases exist concerning the equality of r and r':

1. If $r = r'$, then $f(u,r) \neq f(u',r')$, since $f_r = f_{r'}$ and is injective, so that different inputs give different outputs.

2. If $r \neq r'$, then $f(u,r) \neq f(u',r')$ or $f(u,r) = f(u',r')$. The latter may hold since two distinct injective functions f_r and $f_{r'}$ map on the same output space.

Notice that the problematical second case only happens if $r \neq r'$. Hence, to establish a unique number o, sufficient information about the selection of f has to be attached to its output. The selection is done through choosing the randomizer r. As a consequence, o can be defined as the pair $(f(u,r), r)$. This result gives the following theorem:

Theorem 3.2.1 Let $f : U \times R \to O_f$ be a uniqueness randomization function and $u, u' \in U$, where $u \neq u'$. Then $(f(u,r), r) \neq (f(u',r'), r')$ holds for all $r, r' \in R$.

Proof. The case where $r \neq r'$ obviously guarantees that the pairs $(f(u,r), r)$ and $(f(u',r'), r')$ are distinct. Now consider the case where $r = r'$. Since $u \neq u'$ holds per assumption $f(u,r) \neq f(u',r')$ holds due to the injectivity of f_r and the fact that $r = r'$. Thus, we have $(f(u,r), r) \neq (f(u',r'), r')$ for all $r, r' \in R$. □

This leads to the following raw construction of a unique number o:

$$o := (f(u,r), r), \quad \mathsf{UG}() = u, \quad \mathsf{NG}() = r$$

Obviously, o cannot be truly random because the bits of $f(u,r)$ and r are mutually dependent. Moreover, the output of f at most appears to be random. Thus, the block o at most appears to be random. The quality of randomness depends on how UG, NG and f are chosen. Accordingly, statistical tests seem to be indispensable.

The approach proposed in [Sch01] is a special case of the above construction. There, f is instanced by a symmetric encryption function E_S, where $k = r$ is the corresponding key. Furthermore, u contains a system-wide unique identifier UI concatenated to a counter c, which is incremented in every round. So o is defined as follows in [Sch01]:

$$o = E_S(UI||c||\rho, k)||k$$

where ρ is random padding up to the input block-length. Notice, however, that using concatenation through writing blocks in a row is an unnecessary restriction. This leads to the following section, where techniques to provide increased flexibility are explored.

3.3 Techniques to Provide Flexibility and Efficiency

In the current section, several building blocks are described, which provide flexibility and efficiency for the functions used within collision-free number generators.

3.3.1 Pre-Processing

Several functions match the definition of f. Some of them, such as encryption functions, require that the random input r has a special form. If, for instance, f is the RSA encryption function (cf. Section 2.2.7), then r needs to be a public key (e, n), where n is the modulus and e relatively prime to $\varphi(n)$. Hence, using NG alone does not suffice, since (e, n) is not a pure random string. To remain flexible, it is hence necessary to replace NG with a more powerful process, which uses NG as a subroutine:

Definition 3.3.1 A function Pre, which generates or prepares inputs for one or more functions and uses NG as a subroutine, is called *Pre-Processor*.

The above definition allows the preparation of inputs for even more than one function. This is necessary, because f is not necessarily the only function that needs to have access to NG. For example, UG might also require access to NG for padding.

3.3.2 Efficient Uniqueness Generation

The only task of the uniqueness generator is to output unique numbers. A very efficient method to implement UG is the use of a random counter c concatenated to a system-wide unique identifier UI. UG can obtain the latter through the initialization process. Then, by incrementing c modulo 2^{l_c} in each round, at least $2^{l_c} - 1$ system-wide unique numbers can be established. This gives the following construction of UG:

$$\text{UG} : \emptyset \rightarrow \{0, 1\}^{l_u}, \quad \text{UG}() := UI||c||\rho, \quad c := (c+1) \text{ MOD } 2^{l_c}$$

where $l_u = l_{UI} + l_c + l_\rho$ and ρ some random padding. The counter c can be initialized at random and then incremented modulo 2^{l_c}. In principle, c can also be initialized with a sequence of l_c zeros, but this possibly opens the algorithms to some attacks, since then an adversary knows that *every* uniqueness generator of the system starts with such an initial value.

The advantage of an identifier-based approach is the fact that, apart from the initialization process, no further communication is necessary to guarantee uniqueness. Useful candidates for UI include the IEEE 802 MAC-address of a network device, the Integrated ChipCard Serial Number contained in smartcards, an email-address, a domain-name, a national insurance number, a credit card number, a telephone number, or even a static IP-address. Bear in mind, however, that within the same system all generators have to use the *same type* of unique identifier.

3.3.3 Integration of Blocks

The basic construction given in Section 3.2 left open how the representation of a pair (a, b) is to be implemented in practice. For example, writing a and b in a row can be a solution, but the two blocks can be integrated in any way, as long as injectivity holds. For illustration, the following three techniques are discussed:

1. Static Permutation

2. Dynamic Permutation

3. Adapted Optimal Asymmetric Encryption Padding

The bits of two blocks can be concatenated in any static way, so that (a, b) denotes a bit-permutation over the block $a||b$. Additionally, the result can be expanded by the use of random bits. Then (a, b) denotes a bit-permuted expansion of $a||b$.

When using a dynamic permutation, sufficient information needs to be attached to uniquely identify which permutation has been used. Injectivity holds, because Theorem 3.2.1 can be applied. Unfortunately, such an approach is not useful for bit-permutations, because the permutation-key would be extremely large. A compromise can be the use of dynamic block-permutation or dynamic permuted insertion of blocks. Moreover, static and dynamic permutations can be combined.

A third interesting way to integrate the two blocks is the use of an adapted version of Optimal Asymmetric Encryption Padding (OAEP) [BR95]. This allows a plaintext block to be randomized by a random block and two random oracles (in practice cryptographic hash functions). If we consider a as the plaintext block and b as the random block, then, by use of OAEP, the two blocks result in a block which appears random.

3.3.3.1   Static Permutations

First, a general definition for the concatenation of the bits of n blocks is given.

Definition 3.3.2 Let $s : [0, k-1] \to [0, k-1]$ be a permutation, where $k = \sum_{i=1}^{n} l_{B_i}$ and l_{B_i} the bit-length of block B_i. The *n-Block Bit-Permutation* π_s^n is given by:

$$\pi_s^n : \{0,1\}^{l_{B_1}} \times \ldots \times \{0,1\}^{l_{B_n}} \to \{0,1\}^k, \quad \pi_s^n(B_1, \ldots, B_n) := b_{s(0)} || \ldots || b_{s(k-1)}$$

where b_i denotes the i-th bit of the block $B_1 || \ldots || B_n$.

Such a function is bijective for all inputs if s is static and hence can be applied to intermediate and final results anywhere in the basic construction. However, one can go a step further and define an n-block bit-permuted expansion, where a random block is first attached to the n input-blocks and then a permutation takes places.

Definition 3.3.3 Let $s : [0, k-1] \to [0, k-1]$ be a permutation, where $k = l_r + \sum_{i=1}^{n} l_{B_i}$, l_r the bit-length of the (internally chosen) random expansion and l_{B_i} the bit-length of block B_i. The *n-Block Bit-Permuted Expansion* $\widehat{\pi}_s^n$ is given by:

$$\widehat{\pi}_s^n : \{0,1\}^{l_{B_1}} \times \ldots \times \{0,1\}^{l_{B_n}} \to \{0,1\}^k, \quad \widehat{\pi}_s^n(B_1, \ldots, B_n) := \pi_s^{n+1}(B_1, \ldots, B_n, r)$$

where $r \in_R \{0,1\}^{l_r}$.

Once again, such a function is injective for all inputs if s is static. Moreover, it covers the n-block permutation as a by-product if one sets $l_r = 0$.

3.3.3.2   Dynamic Permutations

If s in not static, then the use of π_s^n or $\widehat{\pi}_s^n$ leads to collisions. This problem can be overcome with the same trick as used in Section 3.2.3 and gives the following corollary:

Corollary 3.3.1 *Let $B, B' \in \{0,1\}^{l_{B_1}} \times \ldots \times \{0,1\}^{l_{B_n}}$, where $B \neq B'$, and $\widehat{\pi}_s^n$ an n-block bit-permuted expansion. Then $(\widehat{\pi}_s^n(B), s) \neq (\widehat{\pi}_{s'}^n(B'), s')$ holds for all s and s'.*

Proof. This directly follows from Theorem 3.2.1, since $\widehat{\pi}_x^n$ is injective for all x. □

Notice that the above corollary also covers n-block bit-permutation. Although an interesting approach, attaching s requires attaching $s(0), \ldots, s(k-1)$ in practice, which in turn gives a very large output. This comes from the fact that a bit-permutation is used. If small blocks of bits are permuted, then the output length can be decreased.

3.3.3.3   Optimal Asymmetric Encryption Padding

In [BR95], a method called Optimal Asymmetric Encryption Padding (OAEP) has been introduced which can be used to turn deterministic encryption schemes into probabilistic ones. The idea is to randomize the plaintext by a random string and use the resulting block as input to the encryption function. The randomization process is bijective, since otherwise one cannot uniquely remove the randomness from the plaintext after the decryption. The intention for using OAEP in the context of the RSA-scheme is to eliminate the malleability property [DDN91] and to achieve security against chosen-ciphertext attacks [Ble98, Sho98].

Now consider the use of OAEP for integrating $f(u, r)$ with r. In the resulting block, the bits of both blocks are well distributed, which is not the case if static or dynamic permutations are used. Since OAEP is bijective, the uniqueness of $(f(u, r), r)$ is preserved and hence, this technique can be used for integration in our basic construction.

Definition 3.3.4 Let $\mathcal{G} : \{0,1\}^{l_b} \rightarrow \{0,1\}^{l_a}$ and $\mathcal{H} : \{0,1\}^{l_a} \rightarrow \{0,1\}^{l_b}$ be two random oracles. *Optimal Asymmetric Encryption Padding* OAEP is defined as follows:

$$\text{OAEP} : \{0,1\}^{l_a} \times \{0,1\}^{l_b} \rightarrow \{0,1\}^{l_a+l_b}, \quad \text{OAEP}(a, b) := a \oplus \mathcal{G}(b) || \mathcal{H}(a \oplus \mathcal{G}(b)) \oplus b$$

Permutation and permuted expansion are both henceforth referred to as *Permutation*. Bit-permutations performed over *one* block B are denoted by $\tau(B)$. Permutation or the use of any injective random oracle over *two* blocks B_1 and B_2 is denoted by $\pi(B_1, B_2)$ and sometimes referred to as *Integration Function*.

3.3.4   Variable Length Uniqueness Randomization

Efficient candidates for f are symmetric block ciphers, where block-lengths are generally static. Typical block-lengths include 64 bits for the Data Encryption Standard (DES) [NIS99], the SKIPJACK-Algorithm [NIS98] and the International Data Encryption Algorithm (IDEA) [LM91], and 128 bits for the Advanced Encryption Standard (AES) [NIS01a]. When using such schemes for f, the block-length is obviously the lower bound for u. Let l_b be the block-length of the considered block-cipher and $l_u > l_b$.

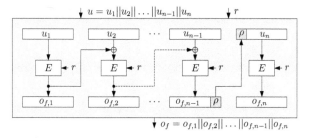

Fig. 3.1: Variable Length Uniqueness Randomization.

Then padding seems to be necessary in UG up to a multiple of l_b. For some applications, this overhead is too large. If, for example, $l_b = 128$ and $l_u = 129$, then the output of f

is 256 bits long, although 129 bits would be sufficient. For this case, a solution known as Ciphertext-Stealing [MM82] can be used. The idea is simple: if $l_u = (n-1)l_b + \varepsilon$, where $\varepsilon < l_b$ and $n > 1$, then the n-th input-block needs to be padded up to l_b bits, that are chosen from the $(n-1)$-st output block of the block-cipher. For an interleaved randomization, it can be interesting to use a variant of the Cipher-Block Chaining Mode [NIS80]. Both techniques are stated together in Figure 3.1.

3.3.5   Variable Final Output-Length

If bits are concatenated to the right side of the output o, then uniqueness is obviously no longer guaranteed. However, on the left side of o blocks of any length can be concatenated without violating the uniqueness. Thereby, one must not exclude the leading zeros of o. The duration of the uniqueness of o, or the expanded block, depends on the length of o. Once all numbers have been generated, uniqueness can no longer be guaranteed if the size of o is not increased. Interestingly, a simple local update, where no communication is necessary, leads to the property that infinitely many system-wide unique numbers can be generated.

The idea is sketched in the following. Assume that a generator has detected that it generated the last number for which uniqueness is guaranteed, e.g. the counter has been increased $2^{l_c} - 1$ times. Then it concatenates a 1-bit to the left side of o and restarts the generation process. For the identifier-based approach, this just means that once again $2^{l_c} - 1$ system-wide unique numbers can be generated. Notice that the new numbers are one bit longer than the numbers generated in the previous setting. If the generator again comes to the point where uniqueness is not guaranteed anymore, it does the same trick, but now with more modifications:

1. The lengths of u and o_f are increased by 1.

2. The permutation or permuted expansion π is adapted appropriately.

3. A 1-bit is concatenated to the left of o.

Step 1 is only possible if suitable candidates for UG and f are used. If u is an identifier concatenated to a counter, one can increase the size of the counter and in the new setting, $2^{l_c+1} - 1$ system-wide unique numbers can be generated. For f, Ciphertext-Stealing can be used to increase the input size. Such a local update can be done without communication between the generators or to a central instance of the system. Number generators, which are in the same or different setting, produce distinct numbers. The case where they are in the same setting is trivial. If they reside in different settings, then the generated numbers have different lengths.

3.4   Three Types of Generators

In the basic approach, both uniqueness and randomness have been taken into account. Furthermore, techniques have been presented which can be used to provide more flexibility. For extended applications, privacy can play an additional role. We assume that u contains information about the instance by which it has been generated. If it is easy to invert f by use of r, then u can be efficiently obtained if o is accessible. Hence, we have to distinguish between applications where o is public (e.g. a pseudonym) and applications where o is secret (e.g. a secret key). If o is secret and f, UG and Pre are chosen appropriately then no further techniques are necessary to hide u. This leads to the first definition of a collision-free number generator (cf. Figure 3.2):

Definition 3.4.1   Let UG $: \emptyset \to U$ be a uniqueness generator, $f : U \times R \to O_f$ a uniqueness randomization function, $\pi : O_f \times R \to O$ an integration function and Pre $: \emptyset \to R$ a pre-processor. The collision-free number generator CFNG1 is given by:

$$\text{CFNG1} : \emptyset \to O, \quad \text{CFNG1}() := \pi(f(u,r),r), \quad \text{UG}() = u, \quad \text{Pre}() = r$$

Corollary 3.4.1   *Let $o = $ CFNG1(). Then o is system-wide unique.*

Proof. This directly follows from Theorem 3.2.1 and the injectivity of π.   □

For applications where o is public, the following three options come to mind:

1. f is a one-way function and r is not sufficient for an efficient inversion of f.

2. f is a one-way function and r is sufficient for an efficient inversion of f, but hidden by use of an injective one-way function g.

3. f is not necessarily a one-way function and r is sufficient for an efficient inversion of f, but $\pi(f(u,r),r)$ is hidden by use of an injective one-way function g.

For the first variant, CFNG1 can be used if f is an asymmetric encryption function and r the associated public key. An example is the use of the RSA encryption function. Here, one has to be careful that the choice of u does not contradict the requirements for secure RSA encryption. Keep in mind that the basic scheme is not a probabilistic encryption scheme, so that using RSA-OAEP should be taken into account.

For the second variant, f could be a symmetric encryption function, where r is the associated secret key. For g, one could use exponentiation in a group where the Discrete Logarithm Problem [McC90] is believed to be hard. If r is rather short, it may be necessary to apply a bit-permuted expansion τ. This leads to the second definition of a collision-free number generator (cf. Figure 3.2):

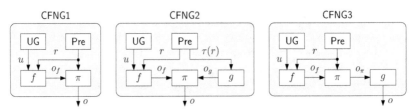

Fig. 3.2: Three Types of Collision-Free Number Generators.

Definition 3.4.2 Let $\mathsf{UG} : \emptyset \to U$ be a uniqueness generator, $f : U \times R \to O_f$ a uniqueness randomization function, $g : O_\tau \to O_g$ an injective one-way function, $\pi : O_f \times O_g \to O$ an integration function, $\tau : R \to O_\tau$ a permutation and $\mathsf{Pre} : \emptyset \to R \times O_\tau$ a pre-processor. The collision-free number generator $\mathsf{CFNG2}$ is given by:

$$\mathsf{CFNG2} : \emptyset \to O, \quad \mathsf{CFNG2}() := \pi(f(u,r), g(\tau(r))), \quad \mathsf{UG}() = u, \quad \mathsf{Pre}() = (r, \tau(r))$$

Corollary 3.4.2 *Let* $o = \mathsf{CFNG2}()$. *Then* o *is system-wide unique.*

Proof. This directly follows from Theorem 3.2.1 and the injectivity of π and $g \circ \tau$. $\square$

For the third variant, f could just be any randomization function that is easy to invert and g again the exponentiation in an appropriate group. This leads to the third definition of a collision-free number generator (cf. Figure 3.2):

Definition 3.4.3 Let $\mathsf{UG} : \emptyset \to U$ a uniqueness generator, $f : U \times R \to O_f$ a uniqueness randomization function, $\pi : O_f \times R \to O_\pi$ an integration function, $g : O_\pi \to O$ an injective one-way function and $\mathsf{Pre} : \emptyset \to R$ a pre-processor. The collision-free number generator $\mathsf{CFNG3}$ is given by:

$$\mathsf{CFNG3} : \emptyset \to O, \quad \mathsf{CFNG3}() := g(\pi(f(u,r), r)), \quad \mathsf{UG}() = u, \quad \mathsf{Pre}() = r$$

Corollary 3.4.3 *Let* $o = \mathsf{CFNG3}()$. *Then* o *is system-wide unique.*

Proof. This directly follows from Theorem 3.2.1 and the injectivity of $g \circ \pi$. $\square$

If g is a one-way function based on a yet unsolved mathematical problem, then the output of $\mathsf{CFNG2}$ is generally larger than that of $\mathsf{CFNG3}$. Furthermore, for two outputs o and o', where $o \neq o'$, of two different runs of $\mathsf{CFNG2}$ it can happen that $r = r'$. Such a collision is publicly detectable, because $g \circ \tau$ is injective. In such a case, the holder of r' (r) can invert f_r ($f_{r'}$). If this is considered to be a problem, then one can use $\mathsf{CFNG3}$ or implement $\mathsf{CFNG2}$ in a tamper-resistant device, such that the holder cannot get r out of the device. Notice that applying g only to r might be useful for scenarios where one has to prove interactively that o has a specific form.

In the remainder of this section we analyze (on an abstract level) whether the requirements for uniqueness, randomness, efficiency and privacy are fulfilled.

Uniqueness: From Theorem 3.2.1 it follows that this requirement is fulfilled for all three types as long as querying UG returns a system-wide unique number.

Randomness: If f is considered as a random oracle, then the output of each generator appears highly random. Notice that this is based solely on intuition.

Efficiency: With respect to communication, all three generators achieve the highest degree of efficiency, since communication is only necessary during initialization. Regarding time, the first generator (if its output is public) is not very fast compared to hash functions or fast pseudo-random number generators. This is the case because asymmetric encryption needs to be used which is time-consuming. The second and third solution give a similar result since g is also slow compared to symmetric schemes. Notice that in Table 3.2 we hence marked all three generator with a time constraint of *medium*. When considering the space of each generator, one has to distinguish between the space of the generation process and the minimal possible length of the generated output. If the output is public and privacy needs to be fulfilled, then CFNG1 achieves the worst result. CFNG2 gives smaller outputs, but is still not satisfactory if used for the generation of Universally Unique Identifiers, for instance. Finally, CFNG3 is the most efficient solution with respect to the length of its output (cf. Section 3.5).

Privacy: It is quite challenging to *prove* the privacy of the three collision-free number generators on such an abstract level. Whether privacy is guaranteed sufficiently, depends on how the various functions are chosen. Notice that using symmetric encryption as a building block complicates a reduction to a computational problem.

3.5   Selected Implementation Issues

The three types of collision-free number generators, defined in Section 3.4, can be instantiated in several ways. For simplicity, we restrict our considerations to a common basis where we only use different settings for f and g, whereas UG, π and τ remain unchanged, aside from varying bit-lengths. For UG, an identifier-based approach is used, where UI is the IEEE 802 MAC-address of the issuer, i.e. $l_{UI} = 48$ bits. To ensure uniqueness for a large number of runs, a 32-bit counter is used. Thus, for u we have a lower bound of $l_u \geq 80$ bits. For π we use ordinary concatenation.

In the following, practical collision-free number generators are discussed with respect to the requirements stated in Section 3.1.2. It is assumed that the output of each generator is publicly available, since this requires special attention to privacy. Table 3.4 contains recommendations for the choice of the parameters for digital signatures in Germany. These include the RSA signature scheme in $\mathbb{Z}_n$ and the Digital Signature Algorithm (DSA) for a cyclic subgroup of $\mathbb{Z}_p^*$ of order q, where p and q are primes,

Algorithm	Parameter	Until 2008	Until 2009	Until 2010	Until 2011
RSA	n	1024	1280	1536	1728
DSA	q	160	160	160	224
	p	1024	1280	1536	2048
EC-DSA	q	180	180	180	224
	p	192	192	192	224

Tab. 3.4: Parameter-Recommendations for Digital Signatures in Germany [PT05].

such that $q|(p-1)$, and for a cyclic subgroup of the Elliptic Curve (EC) [Mil86] group $E(\mathbb{Z}_p)$ of order q, where $q|\#E(\mathbb{Z}_p)$. These parameters are necessary to keep factoring n and to keep computing discrete logarithms with respect to the parameters p and q hard. These recommendations are even of general interest and not just restricted to digital signature schemes. We henceforth use these lower bounds for RSA and Rabin encryption [Rab79] in $\mathbb{Z}_n$, exponentiation in a subgroup of $\mathbb{Z}_p^*$ and scalar multiplication in an EC-subgroup of $E(\mathbb{Z}_p)$. The latter two we use for g and ElGamal.

3.5.1   Examples for CFNG1

To preserve privacy, f must be an injective one-way function, where r is not sufficient for an efficient inversion of f. In practice, f can be instanced by an asymmetric encryption function E_A (cf. Figure 3.3), where r is the public key pk. Compared to symmetric schemes, this generally leads to the drawback that o is rather large. Table 3.5 contains the bit-lengths for the concrete choices of f, that are RSA-OAEP, Rabin-OAEP, ElGamal and EC-ElGamal here.

RSA-OAEP: The use of Plain-RSA is not advisable, since it turned out to suffer from several weaknesses. Thus, we recommend to use RSA-OAEP, so that the input is randomized and then encrypted. Let $n = pq$, where p and q are appropriately chosen primes and $l_n = 1024$. UG, in this case, is designed such that it outputs $u = UI||c||\rho$, where ρ is a 944-bit random padding so that $l_u = 1024$ holds. Furthermore, it needs to be checked if $u < n$, otherwise the uniqueness is not necessarily preserved. For OAEP, one has to choose $\mathcal{G} : \{0,1\}^{944} \to \{0,1\}^{80}$ and $\mathcal{H} : \{0,1\}^{80} \to \{0,1\}^{944}$. We define the encryption function as $E_A((m,r),(e,n)) = c$, where m is the plaintext, r the randomizer, (e,n) the public key and c the ciphertext. The pre-processor is the RSA key generation algorithm (here denoted as KG) and hence outputs $pk = (e,n)$, such that $ed \equiv 1 \pmod{\varphi(n)}$. Accordingly, the function f is defined as follows:

$$f : \{0,1\}^{l_n} \times (\mathbb{Z}_{\varphi(n)}^* \times \{0,1\}^{l_n}) \to \mathbb{Z}_n, \quad f(UI||c||\rho, pk) := E_A((UI||c,\rho), pk)$$

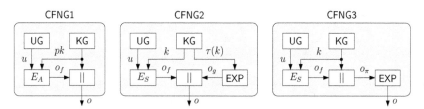

Fig. 3.3: Practical Implementations of Collision-Free Number Generators.

The output o_f of E_A and the corresponding public key pk are then concatenated to each other resulting in the output o. The uniqueness of o is guaranteed by Corollary 3.4.1. Since OAEP has been used, statistical tests [NIS01b] affirmed the conjecture that o provides a high degree of randomness. Notice that RSA-OAEP is a probabilistic scheme, satisfying the requirement for indistinguishability of ciphertexts. Unfortunately, each block consists of 3072 (i.e. $1024 + 2 \cdot 1024$) bits. Obtaining UI (requirement R4.H) is hard if factoring n and computing e-th roots modulo n are hard. Since ciphertexts are indistinguishable, R4.U holds as well.

Rabin-OAEP: Rabin encryption is actually the RSA encryption scheme with the special case $e = 2$. Again, we use OAEP to get a randomized encryption. Since every generator uses $e = 2$, it suffices to concatenate o_f to n, so that each output-block consists of 2048 (i.e. $1024 + 1024$) bits. Privacy can be considered analogous to RSA.

ElGamal: Let $\mathbb{G}_q$ be an appropriate subgroup of $\mathbb{Z}_p^*$. Moreover, let γ be a generator of $\mathbb{G}_q$. We set $u = UI||c||\rho$, with the condition $u < q$. The pre-processor is the key generation algorithm KG and returns the public key $h = \gamma^z$, where $z \in_R \mathbb{Z}_q$. Then f is just an ElGamal encryption which outputs $o_f = (\gamma^r, \gamma^u h^r)$, where γ^u is the plaintext and $r \in_R \mathbb{Z}_q$. Finally, we set $o := o_f||h$. Since encryption and group exponentiation are bijective, uniqueness holds from Corollary 3.4.1. Hiding is fulfilled, since given a ciphertext, h and γ, the problem of finding the plaintext is computationally equivalent to the Diffie-Hellman Problem [DH76]. Regarding the indistinguishability of ciphertexts [GM82], this variant of ElGamal encryption provably reduces to the Decision Diffie-Hellman Problem [Bon98]. Based on the reductions, given o, it is hard to obtain UI or even decide if two numbers on hand have been derived from the same unique identifier. Statistical tests again showed satisfying results regarding the randomness of o. Unfortunately, the output-block is 3072 (i.e. $2 \cdot 1024 + 1024$) bits long.

EC-ElGamal: In this case, $\mathbb{G}_q$ is a cyclic subgroup of $E(\mathbb{Z}_p)$. Since $l_q = 180$ and $l_p = 192$ bits are currently believed to be sufficient [PT05], the bit-length of the output is now 579 (i.e. $2 \cdot (192 + 1) + (192 + 1)$) bits. Notice that we write $192 + 1$ instead of 193, because the x-coordinate of a point lies in $\mathbb{Z}_p$ and the y-coordinate can be coded

E_A	l_u	l_{pk}	l_{o_f}	l_o
RSA-OAEP	1024	2048	1024	3072
Rabin-OAEP	1024	1024	1024	2048
ElGamal	160	1024	2048	3072
EC-ElGamal	180	193	386	579

Tab. 3.5: Bit-Lengths for CFNG1.

uniquely by 1 bit. This technique is known as point-compression and can be found in [IEE00]. Uniqueness, randomness and privacy hold like in the ordinary setting.

3.5.2   Examples for CFNG2

For this variant, it is assumed that f is again a one-way function, but now r is sufficient for an efficient inversion of f. In practice, f can be instanced with a symmetric encryption function E_S (cf. Figure 3.3), where r is the secret key k. So, contrary to using asymmetric schemes, f can be evaluated very fast. However, to fulfill the requirement for privacy, one has to hide k by using a one-way function g. Practical instances for g are the exponentiation (abbreviated by EXP) in a suitable subgroup of $\mathbb{Z}_p^*$ and the scalar multiplication (we again write EXP for simplicity) in a suitable subgroup of $E(\mathbb{Z}_p)$. Table 3.6 contains the bit-lengths for the concrete choices of f, that are DES, IDEA, SKIPJACK and AES here. Sometimes, Ciphertext-Stealing is necessary since the input block-length of DES, IDEA and SKIPJACK is only 64 bits. For the use of AES, random padding is necessary to obtain a block-length of 128 bits.

CS-DES: Here, k needs to be expanded up to 160 bits for exponentiation and up to 180 bits for scalar multiplication. Together with the 80-bit output of E_S, this gives the lengths 1104 (i.e. $80 + 1024$) and 273 (i.e. $80 + 192 + 1$) for o, respectively. The output o contains a DES-ciphertext in plain form. So, a brute force attack can be run which will succeed soon, because the key-length is only 56 bits. Hence, this example fulfills privacy only in the short term.

CS-IDEA and CS-SKIPJACK: For IDEA and SKIPJACK, similar results are achieved. In both cases k needs to be expanded appropriately and the resulting output lengths are 1104 and 273, respectively. As far as privacy is concerned, both variants seem to be more secure since the key-lengths are 128 and 80 bits.

AES: To provide a high level of privacy, it is a good choice to apply the most current encryption standard, that is the AES. Then, however, one gets 1152 (i.e. $128 + 1024$) and 321 (i.e. $128+192+1$) for l_o, which is longer than necessary for several applications.

E_S	l_u	l_k	$l_{\tau(k)}$	$l_{\tau(r)}$ (EC)	l_o	l_o (EC)
CS-DES	80	56	160	180	1104	273
CS-IDEA	80	128	160	180	1104	273
CS-SKIPJACK	80	80	160	180	1104	273
AES	128	128	160	180	1152	321

Tab. 3.6: Bit-Lengths for CFNG2.

Notice that block-ciphers can hardly be proven to be secure under a mathematical assumption. As a consequence, it is also hard to prove how complete the privacy-property is. Intuitively, hiding should hold, because a ciphertext of a block-cipher together with the corresponding key hidden by group-exponentiation can be seen as some kind of hybrid cryptosystem.

3.5.3 Examples for CFNG3

The drawback of CFNG2 is that the output of E_S is accessible in plain form. This enables brute force attacks against the chosen block-cipher. Hence, the use of DES is not recommended. Moreover, the final output-length is still large. An alternative idea is to apply g to the block $E_S(u, k)\|k$. Then, brute force attacks are harder to mount, because one cannot perform equality checks on the output of E_S. Moreover, if E_S is a good mixing-transformation, i.e. realizes the concepts of confusion and diffusion, then $E_S(u, k)\|k$ is highly random. If this is the case, then it seems to be sufficient (in terms of bit-length) to use block-ciphers where the block-length plus the key-length give at least the minimal bit-length required for g. So, contrary to the previous setting, here DES can be a viable choice. The other candidates of f can be treated analogously. We summarize the results for hiding and unlinkability in the following.

Privacy (Hiding): The goal of an attack against the hiding property is to obtain UI, i.e. here the permanent 48-bit MAC-address. To obtain UI, the block u has to be found and then the counter removed. Therefore, o_π has to be accessible. If o_π looks sufficiently random, then the best known attack is Pollard's rho algorithm with a running time of $O(\sqrt{q/2})$. Since we use a block-cipher for f, a sequence of blocks of the form $E_S(u, k)$ appears to be random for a random k. Intuitively, the concatenation of the bits of a random number (here k) and a number that appears to be random (here o_f) results in a block which ultimately appears to be random (here o_π). Again, we performed statistical tests and obtained satisfying results. A different attack could be that the attacker guesses that a specific MAC-address has been used. But then he has to mount a brute force attack to find o_f such that the given number

E_S	l_u	l_k	$l_{\tau(k)}$	$l_{\tau(r)}$ (EC)	l_o	l_o (EC)
CS-DES	80	56	80	100	1024	193
CS-IDEA	80	128	128	128	1024	209
CS-SKIPJACK	80	80	80	100	1024	193
AES	128	128	128	128	1024	257

Tab. 3.7: Bit-Lengths for CFNG3.

is equal to $g(E_S(u, k) \| k)$. Since at least 80 bits (key-length plus padding) are random, this attack is infeasible for a poly-bounded algorithm. Another approach could be to partially invert g in order to obtain the 80 bits of k. So far, no algorithm is known that achieves this attack goal. Intuitively, if an algorithm can obtain *any* selected bit, then it can find all bits and hence break the Discrete Logarithm Problem. With respect to the probability distribution, one can further try to optimize the baby-giant step algorithm [Tes01]. Since the statistical tests did not show any significant difference to a uniform distribution, such an optimization seems to be hardly possible.

Privacy (Unlinkability): For the unlinkability we consider two outputs o and o', generated by the same generator. The corresponding unique values u and u' differ at least in 1 bit. Since f is a mixing-transformation, changing 1 bit of the input results in $l_{o_f}/2$ bit-flips on average for the output o_f. Furthermore, k is chosen independently in every run and so half of the bits will be different on average. Hence, in total, o and o' differ in approximately one half of the bits. Due to the hiding property, UI cannot be efficiently obtained from o or o'.

Table 3.7 summarizes the results regarding the bit-lengths. The shortest practical output-length, while satisfactorily fulfilling the other requirements, is 193 bits. In Section 3.6.1, this efficient variant is illustrated with respect to an application.

3.6   Applications

In the following, the presented techniques are discussed with regard to applications given in Chapter 2. Moreover, further applications are given in the context of privacy.

3.6.1   Universally Unique Identifiers

In RFC 4122, Universally Unique Identifiers have been specified for a length of 128 bits. It is obvious that it is hardly possible to achieve such a short block-length while fulfilling all the requirements stated in Section 3.1.2. As shown in Section 3.5, the shortest length one achieves with our approach is 193 bits. This gives the following two versions that could be taken into account in an extension to RFC 4122.

Identifier-Based Short-Term Privacy: If the privacy of the generated Universally Unique Identifiers only needs to hold for a short period of time (e.g. several days), then we suggest to modify our implementation, such that $l_o = 128$. For instance, if $l_u = 64$ one can use the DES encryption algorithm for E_S, and then set $o = \mathsf{SM}(E_S(u, k)\|k, P)$, where $l_q = l_p = 128$ (here the parity-bits of k would be replaced by random ones), SM the scalar multiplication in the EC-subgroup $\mathbb{G}_q$ and $\langle P \rangle = \mathbb{G}_q$. Pollard's rho algorithm then has a running time of $O(2^{64})$ steps, which is sufficient for a short time.

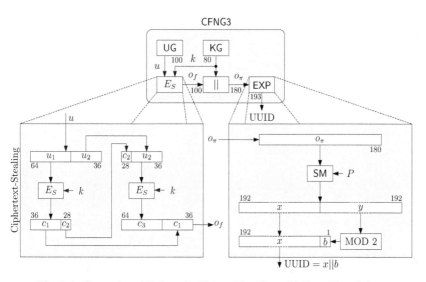

Fig. 3.4: Generation of Universally Unique Identifiers with Long-Term Privacy.

Identifier-Based Long-Term Privacy: If privacy needs to be maintained for a long time (several years), we suggest to use an implementation like the one given in Figure 3.4 with respect to CFNG3 and the bit-lengths recommended in [PT05]. The 100-bit output of UG (e.g. MAC-address with 32-bit counter and padding) is encrypted with SKIPJACK (using Ciphertext-Stealing). Thereby, the left 64 bits are first encrypted using k. The 28 least significant bits of the output are then concatenated to the 36 right bits of u. The resulting block is 64 bits long and also encrypted with k. The output is concatenated to the remainder of the first ciphertext block. Together this gives o_f with a length of 100 bits. Since $l_q = 180$, one has to check if $o_\pi < q$ holds. Otherwise, the encryption process must be repeated. In EXP, o_π is multiplied by P using scalar multiplication SM. The output is a point which consists of two coordinates, each of which lies in $\mathbb{Z}_p$. Using the point-compression technique this gives a 193-bit UUID.

3.6.2 Digital Pseudonyms

In [SS05], we published a solution for user-generated digital pseudonyms. The solution there is equivalent to CFNG1, where f is instanced with the RSA encryption function. Such an approach is interesting for applications where the user is honest and has a strong interest in revealing his identity if necessary. For instance, he wins an auction and wants to prove ownership. An example where the user can be forced to behave honestly is when smartcards are involved. Using CFNG1 for such pseudonyms gives a long pseudonym as output. If this is a problem, then using CFNG2 or CFNG3 could be a viable alternative. If dishonest users are involved then solutions are necessary where a generated pseudonym can be disclosed by a revocation center. Two solutions for this problem are presented and analyzed in Chapter 6.

3.6.3 Remote Controls for Cars

If a remote control for a car is based on symmetric authentication then it suffices to generate the associated key using CFNG1, where f is a block-cipher. This can be done in the factory or by the user after purchase.

Another solution for remote controls is the use of one-time keys, that are generated by a pseudo-random number generator. The car and the key are both initialized with the same seed. If they are used synchronously, then every time the remote control is activated, it generates the next one-time key and sends it to the car. The car does the same and performs an equality check. In practice, it can happen that the remote control and the car get out of sync. For example, one presses the button of the remote control just for fun, although not being within the range of the car. Hence, such schemes have to use some kind of sliding-window technique (e.g. Code Hopping Encoder [Kee03]). To avoid the problem that the pseudo-random number generators generate identical keys (locally and system-wide), it is recommendable to use a collision-free number generator, where NG is a pseudo-random number generator.

3.6.4 ElGamal Keys and Randomizers

To prevent the vulnerabilities stated in Section 2.5.2, we recommend using CFNG1 for the generation of secret keys and randomizers. For instance, one computes the public key $h = \gamma^z$, where $z = \mathsf{CFNG1}()$, $l_z = l_q$, $z < q$ and γ a generator of $\mathbb{G}_q$. If one has white-box access to the generation process, set $h = \mathsf{CFNG3}()$ if g is the group exponentiation in $\mathbb{G}_q$, and obtain the corresponding secret key.

If the Cramer-Shoup cryptosystem is used then it needs to be guaranteed that the public key components are system-wide unique. In Section 2.5.3 we showed that a collision among two public keys does not imply a collision among the corresponding secret keys. Even if secret keys are distinct, the scheme is vulnerable if the public key components collide. Unfortunately, if a public key is chosen to be unique, then the corresponding secret key cannot be computed without solving the Discrete Logarithm Problem. For a public key of the form $h = g_1^{z_1} g_2^{z_2}$, the only way to guarantee uniqueness is if one chooses $dlog_{g_1}(h)$ uniquely. This, however, implies the knowledge of $dlog_{g_1}(g_2)$. Hence, if the standard variant of the Cramer-Shoup cryptosystem is used, then only a trusted party can ensure that public keys are system-wide unique, while being able to compute the corresponding secret keys. Notice that randomizers can be handled in the same way as in the ElGamal cryptosystem.

The design of the simplified variant of the Cramer-Shoup cryptosystem is closely related to the standard ElGamal cryptosystem, i.e. secrets are hidden by just one group exponentiation instead of a multi-exponentiation. Hence, the user can ensure himself that a public key is unique by choosing the corresponding secret key appropriately.

3.6.5 RSA Keys

[Hor98] presented the first approach to ensure that RSA moduli are system-wide unique. The ideas have then been extended in [HS98, HSW98a, HSW98b] and [Sch01]. Using an approach based on master secrets is not necessarily a privacy problem here, since a given modulus does not give opportunities to invert the generation process. Nevertheless, one knows that the same master secret has been used, which might give an advantage, although this is only a conjecture. Similarly to [Sch01], we recommend using CFNG1. As mentioned in Section 2.5.4, it is necessary to ensure uniqueness for RSA-moduli in order to prevent common prime attacks and common modulus attacks. Form the Fundamental Theorem of Arithmetic, every natural number can be written as a product of prime-powers whose representation is unique. If one chooses each prime-factor of an RSA-modulus n system-wide unique, then so is n. The current recommended minimal bit-length for n is 1024 bits (cf. Table 3.4), i.e. the involved primes have a length of 512 bits. Let p and q denote the prime factors of n. Furthermore, let $l_p = l_r + l_{o_f} + l_\rho + 2$, $f : \{0,1\}^{l_u} \times \{0,1\}^{l_r} \to \{0,1\}^{l_{o_f}}$ be a uniqueness randomization function and τ an appropriate static bit-permutation. Then the following abstract steps give a system-wide unique safe-prime p of length l_p:

1. Run $\mathsf{UG}() = u$, choose $r \in_R \{0,1\}^{l_r}$ and compute $o_f = f(u, r)$.

2. Choose $\rho \in_R \{0,1\}^{l_\rho}$.

3. Set $p = 1||\tau(r||o_f||\rho)||1$.

4. If p and $(p-1)/2$ are primes, then return p, otherwise go back to step 2.

If $l_p = 512$, DES can be used for f with Ciphertext-Stealing and $l_u = 80$ bit, and so 430 bits of p are pure random. For more information on this topic we refer to [Sch01].

3.6.6 Untraceable RFID-Tags

RFID-tags are traceable, since they always use the same identifier [Fin99]. This problem could be overcome if a collision-free number generator is used to establish temporary identifiers, that are derived from a permanent one. Using a collision-free number generator has the advantage that the RFID-tags are still *uniquely* identifiable, but are not traceable if a solution is used that ensures unlinkability. Notice, however, that RFID-tags have very little computing resources, i.e. this has to be verified in detail.

3.6.7 Untraceable Smartcards

Here, we have a similar situation to that presented above, but with more advanced resources. For implementing UG, one can use the identifier-based approach of Section 3.3.2, since the Integrated ChipCard Serial Number (ICCSN) is guaranteed to be unique by the manufacturer [RE03]. Beside making a smartcard untraceable, more complex applications can be implemented which use a collision-free number generator as a building block (cf. Chapter 6).

3.6.8 Untraceable Network Devices

Several Internet-applications nowadays have an intrinsic requirement: the anonymity of the user. It is widely known that techniques to achieve anonymity for the application or the transport layer of the TCP/IP-protocol stack are not sufficient to guarantee the anonymity of the user, because he is still identifiable through the network layer (IP-address) and the physical layer (MAC-address). To provide sender- and receiver-anonymity for the network layer, mix-nets can be used [Cha81]. The receiver of IP-packets then does not know the IP-address of the sender, instead, he knows the IP-address of a mixer. Such a technique is useful for end-to-end communications over large distances. The following two scenarios show that the use of the same MAC-address for every communication process enables the traceability of a user.

Location Tracing: The use of mobile technologies has become increasingly prevalent in the recent years. People go online using wireless connections. For example, business people travel around the world by plane and intensively use those technologies. To be able to establish a connection to the Internet, they go online by use of a network device and send a request for a dynamic IP-address. This request contains the IEEE 802 MAC-address of their network device which is then stored in log-files of several servers, such as DHCP-servers or ARP/RARP-servers [Tan03]. By having access to such log-files world-wide, an adversary can obviously trace every network device and hence any traveler using the same mobile device. We call this process *location tracing*.

Session Tracing: For the case that a network device remains in a particular sub-net longer (as it is the case for workstations), sessions established by the corresponding machine are mutually linkable *in* the sub-net. Of course, this does not happen if one eavesdrops the end-to-end channel somewhere else when mixers are used. We call this process *session tracing*. This scenario can especially be a problem, if a user wants to remain anonymous *within* the sub-net, e.g. at a local electronic election.

One idea to make location and session tracing infeasible is to use a new MAC-address *and* a new IP-address every time the network device goes online (or for every session). A new IP-address can be easily obtained by the DHCP-server. For a new MAC-address, however, a generation process is required which runs locally and outputs world-wide unique MAC-addresses. Moreover, MAC-addresses generated by the same instance must not be mutually linkable, since otherwise tracing is once again feasible. These properties can be covered by a collision-free number generator that provides sufficient privacy. Notice that this requires modifications within the related standards.

3.6.9 Unique Passwords

In some situations, it might be important to provide each user of a system with a system-wide unique password. Contrary to giving the freedom of choice to the user, the system itself has to generate the password. Therefore, an appropriate collision-free number generator needs to be run, whose output is mapped to an appropriate alphabet. For instance, one could give a bijective mapping between the binary string and the ASCII-alphabet [ANS86] or a subset of it.

3.6.10 Unique Health Identifiers for Individuals

Unique health identifiers [Uni06], as the name suggests, are required to be unique. An intrinsic requirement is that if just given such an identifier, one must not be able to find any information about the corresponding individual. Thereby, collision-free number generation can be a useful technique. Unfortunately, the output is large compared to common health identifiers. This might be more interesting if smartcards are involved, which is already the case in several countries.

3.7   Concluding Remarks

In this section, the concept of collision-free number generation has been introduced
on an abstract level. Related work has been analyzed and requirements formulated
with respect to the vulnerable applications discussed in Chapter 2. Furthermore, it
has been shown that none of the existing solutions satisfies all requirements. The
basic construction has been shown to be a generalization of the approach in [Sch01].
Based on this core-idea and together with some techniques to provide flexibility, three
types of collision-free number generators have been defined. These approaches have
been given on an abstract level, such that a large range of applications are covered.
Moreover, a special focus has been laid on fulfilling the requirements for privacy and
randomness. Practical examples were then given to show how easy it is to create a
collision-free number generator. Finally, it has been shown where collision-free number
generators can be applied to overcome the vulnerabilities shown in Chapter 2.

An open problem is choosing any of the number generators such that privacy or randomness can be reduced to an unsolved mathematical problem. In other words only
some examples have been shown to fulfill all requirements to sufficiently withstand
attacks of poly-bounded adversaries (cf. RSA and ElGamal variants given in Section 3.5.1). Every time symmetric functions are involved, security cannot (or hardly)
be proven with respect to a problem that is believed to be hard. The only chance
is to follow several design principles and perform statistical evaluations. Although
variants of **CFNG1** can be reduced to the RSA-problem or problems related to the
Discrete Logarithm Problem, such variants are not very practical, since they give a
slow collision-free number generator and extremely large outputs. Nevertheless, they
might be still interesting for systems where the focus lies on other properties, rather
than short bit-lengths (e.g. pseudonym systems).

Basic prototype implementations of the three types of collision-free number generators
have already been provided for JavaCards. For more information we refer to [SS07].

THE CONCEPT OF FUSION

Chapter 4

4.1 Introduction

Let $\mathbb{G}_q$ be a cyclic group of prime order q, where the Discrete Logarithm Problem [McC90] and related problems, such as the Diffie-Hellman Problem [DH76] and the Decision Diffie-Hellman Problem [Bon98], are believed to be hard. In cryptosystems that are based on $\mathbb{G}_q$, (random) secrets are generally hidden by use of exponentiation, i.e. a value $y \in \mathbb{G}_q$ and a generator g are public, but $x = dlog_g(y)$ is kept secret. In general, given y, g and q, computing x is hard under the assumption that the Discrete Logarithm Problem is hard. In Section 2.5, it has been shown that choosing two public key components $y, y' \in \mathbb{G}_q$ to be identical by chance can lead to vulnerabilities in the subsequent cryptosystem. One solution is to use a collision-free number generator to make the key components system-wide unique.

Depending on how a collision-free number generator is implemented, the output can be rather large. If used as an exponent for exponentiation in $\mathbb{G}_q$, this in turn requires that q is chosen sufficiently large, although this is probably not necessary from security point of view. Thus, even if *generated efficiently*, using numbers that have been generated by a collision-free number generator can lead to *inefficient computations*.

Example 4.1.1 The specification requires that the email address of the corresponding instance is used as the unique identifier for UG in CFNG1 (cf. Section 3.4). In [Kle01], the reasonable upper length for email addresses is suggested to be 320 characters (the upper bound assumed for this example). According to [Res01], each of the characters is encoded by 7 bits. Together, this yields the necessary length of 2240 bits for the unique identifier used in UG. If a 32-bit counter is used in UG and Ciphertext-Stealing based on the AES encryption algorithm for f with a 128-bit key, then the output of CFNG1 has a length of 2400 bits. According to Table 3.4 the minimal necessary bitlengths for exponents (i.e. the order of the used group) in discrete-logarithm-based cryptosystems are much shorter. However, to make the secret exponents unique, the order must be sufficiently large such that the output o of CFNG1 is not reduced. Hence, the minimal size of the order of a group in this example is around 2400 bits.

The above example shows that, although the generation of a unique number o is efficient, the use of o as an exponent for discrete-logarithm-based cryptosystems can be quite inefficient because of its length. For such a case this chapter provides the following advantages over the ordinary setting for soft- and hardware solutions:

1. *Software:* In the ordinary setting, the order q would need to be chosen large enough so that no relevant information contained in the secret key gets lost. Our proposal can bring a significant speed-up, since q can be chosen to be smaller without loosing information encoded in the secret key.

2. *Hardware:* If a cryptographic co-processor is used, where q has a fixed length, then l_q might be too short and the relevant information reduced. Thus, in the ordinary setting, the co-processor may not be able to be used. In our setting, if the secret key does not exceed $2l_q$ bits, the hardware can still be used without loosing information encoded in the secret key.

The idea is as follows: a secret exponent x, say a key, is cut into two parts x_1 and x_2 of approximately equal length. If $l_x/2$ is below the lower security bound to keep computing discrete logarithms hard, then x_1 and x_2 are expanded appropriately. The (expanded) key x is now interpreted as a *pair* (x_1, x_2) of integers, instead of as a single integer x. Notice that the key as a pair looks different, but still contains the encoded information. In such a scenario, two alternatives are available:

1. The two components x_1 and x_2 are considered as two separate secret keys (exponents) which are applied one after another.

2. Calculations are carried out over a structure whose members are pairs of the form (x_1, x_2). This, however, requires exponentiations to be done with pairs in the exponent and base, introducing the need for a novel concept of exponentiation that does not simplify the underlying computational problems, i.e. maintains security.

The first solution sounds appropriate but possibly yields security holes. It is not imperative that the application of two keys in a row achieves the goal, namely, that both are equally necessary to perform authorized computations (e.g. encryption or signature generation). Thus, this approach requires an intensive separate analysis for every cryptosystem to which it is applied. For the sake of general applicability we follow the second approach and state the following two minimal requirements.

R1 (Correctness): Cryptosystems work correctly in the new setting.

R2 (Security): Cryptosystems are at least as secure as in the ordinary setting.

Requirement R1 is necessary to ensure that the new setting provides the same set of operations as the ordinary one. For instance, the new kind of exponentiation must follow the basic well-known laws, such as $(g^x)^y = g^{xy}$, $g^x g^y = g^{x+y}$ and $g^x h^x = (gh)^x$. Requirement R2 is necessary to ensure that basic computational problems, such as the Discrete Logarithm Problem, can be defined in the new setting. Furthermore, it needs to be proven that they are at least as hard to solve as in the ordinary setting.

4.1.1   High-Level Description

For $\mathbb{G}_p := \mathbb{G}_q \times \mathbb{G}_q$, with $\cdot$ defined as component-wise multiplication, we define an "exponentiation" where the basis is an element of $\mathbb{G}_p$ and the exponent an element of $\mathbb{F}_p$. Since $p = q^2$, $\mathbb{F}_p$ contains pairs of elements of $\mathbb{Z}_q$. Remarkably, both components of both the basis and the exponent are uniformly included in both components of the resulting element of $\mathbb{G}_p$. This property we call *fusion*. To avoid confusion with ordinary exponentiation, this novel kind of exponentiation is later defined as *fusion-exponentiation* and the problem of inverting it as the *Fusion Discrete Logarithm Problem*. It can be shown that inverting fusion-exponentiation is actually *computationally equivalent* to solving the Discrete Logarithm Problem in $\mathbb{G}_q$. Similarly, but more challengingly, it can be shown that the Fusion Diffie-Hellman Problem is *computationally equivalent* to the Diffie-Hellman Problem in $\mathbb{G}_q$. Moreover, solving the Fusion Decision Diffie-Hellman Problem leads to solving the Decision Diffie-Hellman Problem in $\mathbb{G}_q$. Proving or disproving the converse is currently an open problem.

4.1.2   Contribution and Organization

The contribution of this chapter is the design and realization of the concept of fusion, i.e. the algebraic structure mentioned above. In the next section some necessary preliminaries are given. In Section 4.3, $\mathbb{F}_p$ and $\mathbb{G}_p$ are proven to have the desired properties (i.e. R1 (Correctness)) for the case that $q \equiv 3 \pmod{4}$ and prime. The latter is necessary, otherwise we cannot construct the field $\mathbb{F}_p$. Moreover, by using black-box reductions, several relations between the Discrete Logarithm Problems in $\mathbb{G}_q$ and the Fusion Discrete Logarithm Problems in $\mathbb{G}_p$ are shown (i.e. R2 (Security)). In Section 4.4, the same consideration applies to the case that a group is used whose order is a Blum integer (i.e. the product of two primes congruent 3 modulo 4). This covers schemes that are based on the group of quadratic residues modulo n, where n is the product of two safe-primes. It is argued that the security reductions work there analogously. In Section 4.5, fusion-exponentiation is compared to ordinary exponentiation with particular attention to the computational costs. It is shown that if the exponent is longer than necessary, using the concept of fusion yields a significant speed increase. Finally, concluding remarks are given concerning specialities and open problems.

4.2 Preliminaries

In the following, some basics of number theory and algebra are considered.

4.2.1 Quadratic Residues Modulo a Prime

A well known and important result for this chapter is Fermat's little theorem:

Theorem 4.2.1 (Fermat) *For every integer a, co-prime to an odd prime q, we have:*

$$a^{q-1} \equiv 1 \;(\mathrm{mod}\; q)$$

Using this property one can easily check if a given number $a \in \mathbb{Z}_q^*$ belongs to the set QR_q of quadratic residues or the set QNR_q of quadratic non-residues modulo q.

Proposition 4.2.1 *For $a \in \mathbb{Z}_q^*$, where q is an odd prime, we have:*

$$a^{(q-1)/2} \equiv \begin{cases} 1 \;(\mathrm{mod}\; q) & \text{if } a \in QR_q \\ -1 \;(\mathrm{mod}\; q) & \text{if } a \in QNR_q \end{cases}$$

Based on this proposition the following corollary can be given:

Corollary 4.2.1 *Let q be an odd prime. Then $-1 \notin QR_q$, iff $q \equiv 3 \;(\mathrm{mod}\; 4)$.*

Proof. Let $a \in \mathbb{Z}_q^*$. Then by Proposition 4.2.1 one obtains $a^{(q-1)/2} \equiv \pm 1 \;(\mathrm{mod}\; q)$. Setting $a := -1$ gives $(-1)^{(q-1)/2} \equiv \pm 1 \;(\mathrm{mod}\; q)$. From $q \equiv 3 \;(\mathrm{mod}\; 4)$ it follows that $(q-1)/2$ is odd. So, $(-1)^{(q-1)/2} \equiv -1 \;(\mathrm{mod}\; q)$ and thus $-1 \notin QR_q$. $\qquad\square$

Primes which fulfill $q \equiv 3 \;(\mathrm{mod}\; 4)$ are known to be safe-primes [MVO96]:

Definition 4.2.1 A prime q is called a *safe-prime* if $(q-1)/2$ is an odd prime.

4.2.2 Quadratic Residues Modulo a Composite

Let n be the product of two distinct odd primes p and q. To keep finding the order of $\mathbb{Z}_n^*$ hard, p and q must be chosen such that determining $\varphi(n) = (p-1)(q-1)$ from n is difficult. In order to make the Discrete Logarithm Problem hard in this context, a cyclic subgroup should be used, whose order consists of at least one large prime-factor. Some cryptosystems based on the Discrete Logarithm Problem in cyclic subgroups of $\mathbb{Z}_n^*$ can be found in [Gir91, FO97, Bou00]. Recent approaches, such as [CG98, ACJT00, CL01], are based on QR_n, where n is the product of two safe-primes. Section 4.4 focuses on QR_n for which we need to be able to give a generator g. In the following, we deal with the problem of finding such an element.

Definition 4.2.2 A number $n = pq$ is called a *Blum Integer* if p and q are distinct primes and $p \equiv q \equiv 3 \pmod 4$.

Notice that the product of two safe-primes is always a Blum integer. For such integers, exactly $1/4$ of the elements of $\mathbb{Z}_n^*$ are quadratic residues [MVO96]:

Proposition 4.2.2 *Let n be a Blum integer. Then $|QR_n| = \frac{\varphi(n)}{4}$.*

When using multiplicative groups modulo a natural number n, it is often important to find elements that can be used to generate the whole group. The following fundamental result is useful to decide whether such generators modulo a number n exist.

Theorem 4.2.2 (Gauss) *Generators modulo a natural number n exist, if and only if $n \in \{1, 2, 4, p^\alpha, 2p^\alpha\}$, where p is an odd prime and α a natural number.*

From this theorem it directly follows that $\mathbb{Z}_n^*$ is not a cyclic group, because $n = pq$.

Theorem 4.2.3 (Lagrange) *The order of a subgroup divides the order of the group.*

The converse of Lagrange's theorem does not necessarily hold. The following theorem at least gives a partial result:

Theorem 4.2.4 (Cauchy) *Let $\mathbb{G}_m$ be a group of order m and p a prime divisor of m. Then at least one element $a \in \mathbb{G}_m$ exists such that $ord(a) = p$.*

By assumption, p and q are distinct safe-primes, i.e. there exist two further distinct odd primes p' and q', such that $p = 2p' + 1$ and $q = 2q' + 1$. The order of $\mathbb{Z}_n^*$ is hence $\varphi(n) = (p-1)(q-1) = 4p'q'$. Thus for $\mathbb{Z}_n^*$, the above theorem indicates that *at least* cyclic subgroups of order 2, p' and q' exist. However, further subgroups *may* even exist whose order divides $\varphi(n)$, but this is not necessarily the case and requires attention.

Lemma 4.2.1 *Let $a, b \in \mathbb{G}$, where $\mathbb{G}$ is abelian. Then $ord(ab) = lcm(ord(a), ord(b))$.*

Proof. Computing $(ab)^i$ results in the neutral element (i.e. gives a cycle) if and only if $i \equiv 0 \pmod{ord(a)}$ and $i \equiv 0 \pmod{ord(b)}$. This obviously only happens if $i = 0$ or $i = lcm(ord(a), ord(b))$ holds. $\square$

Based on the above considerations one can show the validity of the following:

Proposition 4.2.3 *Let $a \in \mathbb{Z}_n^*$, where n is the product of two distinct safe-primes $p = 2p' + 1$ and $q = 2q' + 1$. Then $ord(a) \in \{1, 2, p', q', 2p', 2q', p'q', 2p'q'\}$.*

Proof. From Theorem 4.2.2 it follows that $\mathbb{Z}_n^*$ is not a cyclic group. So, there exists no element of order $\varphi(n) = 4p'q'$. It remains to consider every divisor of $4p'q'$, i.e. every number of the set $D = \{1, 2, 4, p', q', 2p', 2q', 4p', 4q', p'q', 2p'q'\}$. Notice that $ord(a) = 1$ is trivial. Let $r \in D \setminus \{1\}$ and prime. From Theorem 4.2.4 is follows that at least one element in $\mathbb{Z}_n^*$ of order r must exist. Therefore, elements with order 2, p' and q' must also exist. From Lemma 4.2.1 it follows that given elements of order 2, p' and q', one can establish elements of order $2p'$, $2q'$, $p'q'$ and $2p'q'$. The remaining candidates in D that need to be considered are 4, $4p'$ and $4q'$. Assume that an element of order 4 exists. Then by Lemma 4.2.1 one can establish elements of order $4p'$, $4q'$ and $4p'q'$. The latter, as mentioned, produces a contradiction through Theorem 4.2.2. Thus, no elements exist of order 4 or a multiple of it. $\square$

Lemma 4.2.2 *Let n and a be defined as in Proposition 4.2.3. Then $ord(a) \in \{p'q', 2p'q'\}$, iff $gcd(a \pm 1, n) = 1$.*

Proof. From Proposition 4.2.3 it follows that for an $a \in \mathbb{Z}_n^*$, the possible orders are 1, 2, p', q', $2p'$, $2q'$, $p'q'$ and $2p'q'$. There is, however, the special restriction $gcd(a \pm 1, n) = 1$. If $ord(a) = 1$, then $a - 1$ is a multiple of n which contradicts the assumption that $gcd(a \pm 1, n) = 1$. If $ord(a) = 2$ then $a^2 \equiv 1 \pmod{n}$ and hence $a^2 - 1$ is a multiple of n. Since $a^2 - 1 = (a+1)(a-1)$ holds, it follows that $gcd(a + 1, n) \neq 1$ or $gcd(a - 1, n) \neq 1$ must hold, which again contradicts the assumption that $gcd(a \pm 1, n) = 1$. Now consider $ord(a) = q'$, without loosing generality. Then $a^{q'} \equiv 1 \pmod{n}$ must hold and hence so $a^{q'} \equiv 1 \pmod{p}$. Thus, if $a \equiv 1 \pmod{p}$ it follows that p divides $a - 1$ which contradicts the assumption. For all other values of a, $a^{q'} \equiv 1 \pmod{p}$ yields a contradiction because q' does not divide $\varphi(p) = 2p'$. For $ord(a) = 2q'$ one gets $a^{2q'} \equiv 1 \pmod{n}$ and thus $(a^2)^{q'} \equiv 1 \pmod{p}$. If $a^2 \equiv 1 \pmod{p}$, then p must be a divisor of $a^2 - 1$. This again contradicts the assumption since this implies that p either divides $a + 1$ or $a - 1$. For all other instances of a^2, $(a^2)^{q'} \equiv 1 \pmod{p}$ yields a contradiction, since q' does not divide $\varphi(p) = 2p'$. If $ord(a) = p'q'$ then $a^{p'q'} \equiv 1 \pmod{n}$ and thus $a^{p'q'} \equiv 1 \pmod{p}$ and $a^{p'q'} \equiv 1 \pmod{q}$. Obviously both hold. Hence, $\langle a \rangle = QR_n$ must hold for such an a, since by Proposition 4.2.2, $|QR_n| = \frac{\varphi(n)}{4} = p'q'$. Finally, let $ord(a) = 2p'q'$. This only holds if $(a^2)^{p'q'} \equiv 1 \pmod{n}$ which only holds if $1 \in QR_p$ and $1 \in QR_q$. This is, however, the case because n is a Blum integer, i.e. $p \equiv q \equiv 3 \pmod{4}$. $\square$

Corollary 4.2.2 *Let n and a be defined as in Proposition 4.2.3, $g = a^2$ and $gcd(a \pm 1, n) = 1$. Then g is a generator of QR_n.*

Proof. By Proposition 4.2.2, the number of quadratic residues modulo a Blum integer n is exactly $\frac{\varphi(n)}{4}$. Since the modulus n used here is such an integer, the order of QR_n is $\frac{\varphi(n)}{4} = \frac{4p'q'}{4} = p'q'$. Setting $g = a^2$, the order of g must divide $p'q'$, since $g \in QR_n$. By Lemma 4.2.2, $ord(a) \neq 2p'$ and hence $ord(a^2) \neq p'$. Analogously, $ord(a^2) \neq q'$. As $gcd(a \pm 1, n) = 1$, $a^2 \not\equiv 1 \pmod{n}$ holds and thus $ord(g) = p'q'$. $\square$

The above results can also be found in [CG98] and [ACJT00], but without any details. Henceforth, let $\tilde{n} = p'q'$. When doing exponentiations in QR_n, the exponent generally needs to be chosen smaller than $\tilde{n}$. Since $\tilde{n}$ is normally unknown, at least the bit-length $l_{\tilde{n}}$ needs to be public so that an exponent can be chosen smaller than

$\widetilde{n}$. When performing exponentiations where the basis is a generator g of the group, it is important that the resulting element is also a generator or has a large order such that computing discrete logarithms is infeasible. In groups of prime order this is the case for all elements except 1. In the following proposition it is shown that choosing an element z from $\mathbb{Z}_{\widetilde{n}}$, such that the order of g^z is 1, p' or q', happens with negligible probability. For simplicity, we consider $z \in \mathbb{Z}_{\widetilde{n}}$ instead of $z \in \{0,1\}^{l_{\widetilde{n}}-1}$ (which would have been the case in practice), since this only minimally effects the estimation.

Proposition 4.2.4 *Let g be a generator of QR_n and z an integer chosen at random from $\mathbb{Z}_{\widetilde{n}}$. Then the probability that $ord(g^z) \neq \widetilde{n}$ is negligible.*

Proof. Since g is a generator of QR_n, it suffices to show that $\Pr[gcd(z, \widetilde{n}) \neq 1]$ is negligible. The probability that z is such an element is

$$1 - \frac{\varphi(\widetilde{n})}{\widetilde{n}} = 1 - \frac{(p'-1)(q'-1)}{p'q'} = 1 - \frac{p'q' - p' - q' + 1}{p'q'} = \frac{p' + q' - 1}{p'q'}$$

Without loosing generality let $p' < q'$. In the following, p' is taken as the lower bound for q' in the last term and -1 is neglected in the numerator:

$$\frac{p' + q' - 1}{p'q'} < \frac{p' + p'}{p'p'} = \frac{2p'}{p'^2} = \frac{2}{p'}$$

which is negligible, since $l_{p'} \approx l_{\widetilde{n}}/2$. $\qquad\qquad\qquad\qquad\qquad\qquad\qquad\qquad\qquad\qquad\Box$

Notice, that if by accident one chooses z such that $gcd(z, \widetilde{n}) \neq 1$, then he can efficiently factor n. This would produce a security hole if the order is required to be unknown.

Interactive proofs for the knowledge or equality of discrete logarithms generally require the knowledge of the group order. There exist approaches which by-pass this problem for groups of unknown order by performing some computations in $\mathbb{Z}$, i.e. without modulo reduction [SV97, GSV98]. If designed properly, such interactive proofs are called *statistically zero-knowledge*. To achieve a proof that is statistically zero-knowledge, exponents are chosen to be a lot larger than the order of the group. For exponents that are required to be system-wide unique, as it is the case in this work, exponents larger than and not co-prime to $\widetilde{n}$ must not be permitted. Otherwise, the uniqueness-property disappears due to implicit modulo reduction.

We henceforth assume that computations in $\mathbb{Z}_{\widetilde{n}}$ are done modulo $\widetilde{n}$ and computations in QR_n are done modulo n. Thus, we do not write MOD $\widetilde{n}$ or MOD n for simplicity.

4.2.3   The Discrete Logarithm Problem Family

The Discrete Logarithm Problem [McC90], the Diffie-Hellman Problem [DH76] and the Decision Diffie-Hellman Problem [Bon98] are widely adopted in the design of cryptosystems, such as [ElG85a, CS98]. In the following, they are defined for $\mathbb{G}_q$.

Definition 4.2.3 Let $\mathbb{G}_q$ be a group of prime order q and $g \in \mathbb{G}_q \setminus \{1\}$.

1. Let $y = g^x$, where $x \in \mathbb{Z}_q$. The *Discrete Logarithm Problem* (DLP) is the following: given y, g and q, find x.

2. Let $y_1 = g^{x_1}$ and $y_2 = g^{x_2}$, where $x_1, x_2 \in \mathbb{Z}_q$. The *Diffie-Hellman Problem* (DHP) is the following: given y_1, y_2, g and q, compute $y_3 = g^{x_1 x_2}$.

3. Let $y_1 = g^{x_1}$, $y_2 = g^{x_2}$ and $y_3 = g^{x_3}$, where $x_1, x_2, x_3 \in \mathbb{Z}_q$. The *Decision Diffie-Hellman Problem* (DDP) is the following: given y_1, y_2, y_3, g and q, decide if $x_3 \equiv x_1 x_2 \pmod{q}$ holds.

Groups where these problems are believed to be hard include [Bon98]:

- The cyclic subgroup of $\mathbb{Z}_{p'}^*$[1] of prime order q, where $q|(p'-1)$ and $q > p'^{1/10}$.

- The cyclic subgroup of $E(\mathbb{Z}_{p'})$ of prime order q, where $q|\#E(\mathbb{Z}_{p'})$.

- The cyclic group QR_n, where n is the product of two safe-primes.

Apart from the fundamental problems listed above, solutions can be found based on related problems including the Matching Diffie-Hellman Problem [FTY96, HTY99], the Square-Exponent (or Squaring) Diffie-Hellman Problem [MW96, BDS98, Wol99], the Inverse Exponent (or Divisible) Diffie-Hellman Problem [PS00, BDZ03] and the Gap Diffie-Hellman Problem [OP01]. Some results considering relations between several of these problems can be found in [Mau94, Sho97, MW98b, MW98a, MW99, Wol99, MW00] and [Kil01]. Most of the results stated above assume that discrete logarithms are *uniformly* distributed. Definitions and studies concerning *non-uniform* distributions can be found in [BT00, Tes01] and in Section 5.2.1.

For the proof of Theorem 4.3.6 we need the Divisible Diffie-Hellman Problem [BDZ03]:

Definition 4.2.4 Let $\mathbb{G}_q$ be a group of prime order q, $g \in \mathbb{G}_q \setminus \{1\}$, $y_1 = g^{x_1}$ and $y_2 = g^{x_2}$, where $x_1, x_2 \in \mathbb{Z}_q$. The *Divisible Diffie-Hellman Problem* (Div-DHP) is the following: given y_1, y_2, g and q, compute $y_3 = g^{x_1 x_2^{-1}}$.

4.2.4 Black-Box Reductions

When analyzing relations between computational problems, oracles, which are able to solve a desired problem in poly-time (or zero-time), are used as a sub-routine. This strong technique is also known as *black-box reduction*, since one does not care about how an oracles solves a problem, but just that it solves it. Throughout this work,

---

[1] We use p' instead of p since we later set $p = q^2$.

an oracle solving a problem P is denoted by $\mathcal{O}_P$. A widely used notation for the reduction of problem A to problem B is A $\leq_p$ B, which means that solving B leads to a solution of A with poly-bounded additional costs in terms of time and space (p stands for polynomial). If A $\leq_p$ B and B $\leq_p$ A both hold, then A and B are said to be computationally equivalent [MVO96]. In that case, one often writes A $\equiv_p$ B. For several proofs in this chapter, the following oracles are used as building blocks:

Definition 4.2.5 Let $\mathbb{G}_q$ be a group of prime order q, $g \in \mathbb{G}_q \setminus \{1\}$, $y = g^x$, $y_1 = g^{x_1}$, $y_2 = g^{x_2}$, where $x, x_1, x_2 \in \mathbb{Z}_q$. The oracles for solving the DLP, DHP and Div-DHP are $\mathcal{O}_{\text{DLP}}(y, g) = x$, $\mathcal{O}_{\text{DHP}}(y_1, y_2, g) = g^{x_1 x_2}$ and $\mathcal{O}_{\text{Div-DHP}}(y_1, y_2, g) = g^{x_1 x_2^{-1}}$.

Proposition 4.2.5 Div-DHP $\equiv_p$ DHP

Proof. Let $g \in \mathbb{G}_q \setminus \{1\}$, $y_1 = g^{x_1}$ and $y_2 = g^{x_2}$, where $x_1, x_2 \in \mathbb{Z}_q$. It has to be shown that (1) Div-DHP $\leq_p$ DHP and (2) DHP $\leq_p$ Div-DHP.

(1) $g^{x_2^{-1}}$ can be obtained using $\mathcal{O}_{\text{DHP}}$ as follows:

$$\mathcal{O}_{\text{DHP}}(g, g, y_2) = \mathcal{O}_{\text{DHP}}(y_2^{x_2^{-1}}, y_2^{x_2^{-1}}, y_2) = y_2^{x_2^{-1} x_2^{-1}} = (g^{x_2})^{x_2^{-2}} = g^{x_2^{-1}}$$

Then $y_3 = g^{x_1 x_2^{-1}}$ can be obtained using $\mathcal{O}_{\text{DHP}}(g^{x_2^{-1}}, y_1, g)$.

(2) $g^{x_1 x_2}$ can be obtained using $\mathcal{O}_{\text{Div-DHP}}$ as follows:

$$\mathcal{O}_{\text{Div-DHP}}(y_1, g, y_2) = \mathcal{O}_{\text{Div-DHP}}(y_2^{x_2^{-1} x_1}, y_2^{x_2^{-1}}, y_2) = y_2^{x_2^{-1} x_1 (x_2^{-1})^{-1}} = (g^{x_2})^{x_1} = g^{x_1 x_2}$$

Since the above steps require negligible additional costs, Div-DHP $\equiv_p$ DHP holds. $\square$

4.3   Fusion based on a Group of Prime Order

The goal of this section is to design a special kind of exponentiation where "exponents" are elements of $\mathbb{Z}_q^2$ instead of just integers. For security reasons each component of the "exponent" must be uniformly contained in the computed results. We call this property *fusion*, since each component should have an equally strong influence on the output of any calculation. The following subsection briefly sketches the basic properties of ordinary exponentiation which are necessary to define a function ξ, later referred to as *fusion-exponentiation*.

4.3.1   Ordinary Exponentiation

Cryptosystems based on the Discrete Logarithm Problem [McC90] are often defined over a group of prime order. On the one hand, this has the advantage that every exponent other than 0 has a multiplicative inverse modulo the group order. On the other hand, no non-trivial subgroups may exist which could enable attacks (such as the Pohlig-Hellman algorithm [PH78]) to solve the Discrete Logarithm Problem faster than with standard techniques (such as Pollard's rho algorithm [Pol78]).

Let $\mathbb{G}_q$ be a (multiplicative) group of prime order q. An element $a \in \mathbb{G}_q$ may be raised to the power of $n \in \mathbb{Z}$ (in particular $n \in \mathbb{Z}_q$) by computing:

$$a^n := \underbrace{a \cdot \ldots \cdot a}_{n\text{-times}}$$

This operation is called *exponentiation*, where a is the *basis* and n the *exponent*. To be able to simply compare (ordinary) exponentiation (as defined above) with the novel kind of "exponentiation" described in this section, we define:

$$\zeta : \mathbb{G}_q \times \mathbb{Z}_q \to \mathbb{G}_q, \quad \zeta(a, n) := a^n$$

For all $a, b \in \mathbb{G}_q$ and $n, m \in \mathbb{Z}_q$ exponentiation follows the following laws:

$$(a^n)^m = a^{nm}, \quad a^{n+m} = a^n a^m, \quad (ab)^n = a^n b^n$$

Written using ζ this gives:

$$\zeta(\zeta(a, n), m) = \zeta(a, nm) \tag{4.3.1}$$

$$\zeta(a, n + m) = \zeta(a, n)\zeta(a, m) \tag{4.3.2}$$

$$\zeta(ab, n) = \zeta(a, n)\zeta(b, n) \tag{4.3.3}$$

As a by-product one obtains:

$$a^0 = 1 \quad \text{and} \quad a^{-n} = (a^n)^{-1} \tag{4.3.4}$$

It is quite obvious that (4.3.2) (resp. (4.3.3)) is a group homomorphism for any fixed $a \in \mathbb{G}_q$ (resp. $n \in \mathbb{Z}_q$). The elementary properties stated above are necessary to realize (basic) discrete-logarithm-based cryptosystems.

4.3.2 Exponents that are Pairs of Integers

In the fusion-setting exponents are required to be pairs of integers of $\mathbb{Z}_q$. In a group $\mathbb{G}_q$ of order q it is essential that the product c of two random exponents $a, b \in \mathbb{Z}_q$ gives no information about a and b. In the ordinary setting this is trivial, since $c \equiv ab \pmod{q}$. In the fusion-setting, however, one has two exponents of the form $(a, b), (c, d) \in \mathbb{Z}_q^2$ and computes the product $(e, f) = (a, b)(c, d)$. Thereby, it is important that $\cdot$ is defined such that e and f contain uniform amounts of information about a, b, c and d. This leads to the necessity of the following theorem:

Theorem 4.3.1 *Let q be a prime. $\mathbb{Z}_q[X]/(X^2+1)$ is a field, iff $q \equiv 3 \pmod 4$.*

Proof. Since $\mathbb{Z}_q[X]$ is a polynomial ring over the field $\mathbb{Z}_q$, it suffices to show that (X^2+1) is a maximal ideal. This is the case if X^2+1 is irreducible in $\mathbb{Z}_q$ because $\mathbb{Z}_q[X]$ is a principle ideal domain. Since X^2+1 is a 2-degree polynomial, irreducibility is achieved by ensuring that $X^2+1 \not\equiv 0 \pmod q$ holds for all $X \in \mathbb{Z}_q$. This is equivalent to -1 being a quadratic non-residue modulo q. By Euler's criterion $(-1)^{(q-1)/2} \not\equiv 1 \pmod q$ holds, if and only if $(q-1)/2$ is odd which is equivalent to $q \equiv 3 \pmod 4$. $\qquad\square$

Henceforth $\mathbb{F}_p = \mathbb{Z}_q[X]/(X^2+1)$, where $p = q^2$ and $q \equiv 3 \pmod 4$, prime and known. One consequence of the above theorem is that $+$ and $\cdot$ are computed in a similar fashion to complex numbers. Let $(a,b),(c,d) \in \mathbb{F}_p$. Then

$$(a,b) + (c,d) = (a+c, b+d) \tag{4.3.5}$$

which is equivalent to $(a+ib) + (c+id) = (a+c) + i(b+d)$ if (a,b) and (c,d) are interpreted as Gaussian integers. Furthermore,

$$(a,b)(c,d) = (ac - bd, ad + bc) \tag{4.3.6}$$

which is equivalent to $(a+ib)(c+id) = (ac - bd) + i(ad + bc)$ if (a,b) and (c,d) are once again interpreted as Gaussian integers.

4.3.3 Fusion-Exponentiation

This subsection presents a particular function ξ, which shares basic properties with ordinary exponentiation ζ, but where "bases" are elements of $\mathbb{G}_p$ and "exponents" are elements of $\mathbb{F}_p$. Thereby, $\mathbb{G}_p := \mathbb{G}_q \times \mathbb{G}_q$, where $\cdot$ is defined component-wise:

$$(a,b)(c,d) = (ac, bd) \tag{4.3.7}$$

for all $a,b,c,d \in \mathbb{G}_q$. Clearly, $\mathbb{G}_p$ is a group of order $p = q^2$. For simplicity, we sometimes denote elements (a_1, a_2) of $\mathbb{F}_p$ or of $\mathbb{G}_p$ by sans-serif font letters, i.e. $\mathsf{a} = (a_1, a_2)$. ξ is now defined as follows:

$$\xi : \mathbb{G}_p \times \mathbb{F}_p \to \mathbb{G}_p, \quad \xi(\mathsf{a},\mathsf{n}) := (a_1^{n_1} a_2^{-n_2}, a_1^{n_2} a_2^{n_1}) \tag{4.3.8}$$

Furthermore, for fixed $\mathsf{a} \in \mathbb{G}_p$, the mapping $\xi_\mathsf{a} : \mathbb{F}_p \to \mathbb{G}_p$ is defined by $\xi_\mathsf{a}(\mathsf{n}) := \xi(\mathsf{a},\mathsf{n})$, and similarly for fixed $\mathsf{n} \in \mathbb{F}_p$, the mapping $\xi_\mathsf{n} : \mathbb{G}_p \to \mathbb{G}_p$ is given by $\xi_\mathsf{n}(\mathsf{a}) := \xi(\mathsf{a},\mathsf{n})$. In the following it is shown that for ξ the same properties hold as given in (4.3.1), (4.3.2) and (4.3.3) for ζ, i.e.:

$$\xi(\xi(\mathsf{a},\mathsf{n}),\mathsf{m}) = \xi(\mathsf{a},\mathsf{nm}) \tag{4.3.9}$$

$$\xi(\mathsf{a},\mathsf{n}+\mathsf{m}) = \xi(\mathsf{a},\mathsf{n})\xi(\mathsf{a},\mathsf{m}) \tag{4.3.10}$$

$$\xi(\mathsf{ab},\mathsf{n}) = \xi(\mathsf{a},\mathsf{n})\xi(\mathsf{b},\mathsf{n}) \tag{4.3.11}$$

Theorem 4.3.2 *Let* $a, b \in \mathbb{G}_p$, $n, m \in \mathbb{F}_p$. *Then (4.3.9), (4.3.10) and (4.3.11) hold.*

Proof. For simplicity, we set $a = (a, b)$, $b = (c, d)$, $n = (e, f)$ and $m = (g, h)$. For (4.3.9) one has to show that the following holds:

$$\xi(\xi((a, b), (e, f)), (g, h)) = \xi((a, b), (e, f)(g, h))$$

From the left hand side:

$$
\begin{aligned}
\xi(\xi((a, b), (e, f)), (g, h)) &\overset{(4.3.8)}{=} \xi((a^e b^{-f}, a^f b^e), (g, h)) \\
&\overset{(4.3.8)}{=} ((a^e b^{-f})^g (a^f b^e)^{-h}, (a^e b^{-f})^h (a^f b^e)^g) \\
&\overset{(4.3.3)}{=} (a^{eg} b^{-fg} a^{-fh} b^{-eh}, a^{eh} b^{-fh} a^{fg} b^{eg}) \\
&\overset{(4.3.2)}{=} (a^{eg-fh} b^{-fg-eh}, a^{eh+fg} b^{-fh+eg}) \\
&= (a^{eg-fh} b^{-eh-fg}, a^{eh+fg} b^{eg-fh}) \\
&\overset{(4.3.8)}{=} \xi((a, b), (eg - fh, eh + fg)) \\
&\overset{(4.3.6)}{=} \xi((a, b), (e, f)(g, h))
\end{aligned}
$$

(4.3.10) requires that $\xi_{(a,b)}$ is a group homomorphism $\mathbb{F}_p \to \mathbb{G}_p$, i.e.:

$$\xi((a, b), (e, f) + (g, h)) = \xi((a, b), (e, f))\xi((a, b), (g, h))$$

Again from the left hand side:

$$
\begin{aligned}
\xi((a, b), (e, f) + (g, h)) &\overset{(4.3.5)}{=} \xi((a, b), (e + g, f + h)) \\
&\overset{(4.3.8)}{=} (a^{e+g} b^{-f-h}, a^{f+h} b^{e+g}) \\
&\overset{(4.3.2)}{=} (a^e b^{-f} a^g b^{-h}, a^f b^e a^h b^g) \\
&\overset{(4.3.7)}{=} (a^e b^{-f}, a^f b^e)(a^g b^{-h}, a^h b^g) \\
&\overset{(4.3.8)}{=} \xi((a, b), (e, f))\xi((a, b), (g, h))
\end{aligned}
$$

(4.3.11) requires that $\xi_{(e,f)}$ is a group homomorphism $\mathbb{G}_p \to \mathbb{G}_p$, i.e.:

$$\xi((a, b)(c, d), (e, f)) = \xi((a, b), (e, f))\xi((c, d), (e, f))$$

Again from the left hand side:

$$
\begin{aligned}
\xi((a, b)(c, d), (e, f)) &\overset{(4.3.7)}{=} \xi((ac, bd), (e, f)) \\
&\overset{(4.3.8)}{=} ((ac)^e (bd)^{-f}, (ac)^f (bd)^e) \\
&\overset{(4.3.3)}{=} (a^e c^e b^{-f} d^{-f}, a^f c^f b^e d^e) \\
&\overset{(4.3.7)}{=} (a^e b^{-f}, a^f b^e)(c^e d^{-f}, c^f d^e) \\
&\overset{(4.3.8)}{=} \xi((a, b), (e, f))\xi((c, d), (e, f))
\end{aligned}
$$

$\square$

Section 4.3.5 gives some proofs related to the difficulty of inverting $\xi_{(a,b)}$ for any fixed $(a, b) \in \mathbb{G}_p$. Since ξ has the same basic properties as ordinary exponentiation, the following is defined as *Fusion-Exponentiation*:

$$\mathsf{a}^{\mathsf{n}} := \xi(\mathsf{a}, \mathsf{n}) \quad \text{resp.} \quad (a, b)^{(e,f)} := \xi((a, b), (e, f))$$

An important consequence is that (cf. (4.3.4)) $(a, b)^{(0,0)} = (1, 1)$ and

$$(a, b)^{-(e,f)} = (a^{-e}b^f, a^{-f}b^{-e}) = ((a^e b^{-f})^{-1}, (a^f b^e)^{-1}) = ((a, b)^{(e,f)})^{-1}$$

where $(0, 0)$ (resp. $(1, 1)$) is the neutral element of $\mathbb{F}_p$ (resp. $\mathbb{G}_p$).

4.3.4 Generating Bases

To be able to use fusion-exponentiation for discrete-logarithm-based cryptosystems, one has to show that elements $(a, b) \in \mathbb{G}_p$ of (sufficiently) large "order" exist. Since $p = q^2$, it can be shown that each element of $\mathbb{G}_p$ except $(1, 1)$ has "order" p with respect to fusion-exponentiation.

Theorem 4.3.3 $\mathbb{G}_p = \{(a, b)^{(c,d)} \mid (a, b) \in \mathbb{G}_p \setminus \{(1, 1)\}, (c, d) \in \mathbb{F}_p\}.$

Proof. Since $|\mathbb{F}_p| = |\mathbb{G}_p|$ it suffices to show that fusion-exponentiation is injective. Let $a \in \mathbb{G}_q \setminus \{1\}$ and $b \in \mathbb{G}_q$. Since q is an odd prime one can set $w = dlog_a(b)$. Let $(c, d), (e, f) \in \mathbb{F}_p$, where $(c, d) \neq (e, f)$. Then

$$\begin{aligned} (a, b)^{(c,d)} &= (a^c b^{-d}, a^d b^c) = (a^{c-wd}, a^{d+wc}), \\ (a, b)^{(e,f)} &= (a^e b^{-f}, a^f b^e) = (a^{e-wf}, a^{f+we}). \end{aligned}$$

If $c = e$ then $d = f$ needs to hold to satisfy $(a, b)^{(c,d)} = (a, b)^{(e,f)}$, which contradicts the assumption $(c, d) \neq (e, f)$. Similarly, $d = f$ implies $c = e$, thus we can assume $c \neq e$ and $d \neq f$. We have $c - wd \neq e - wf$ or $c - wd = e - wf$. The first case leads to $(a, b)^{(c,d)} \neq (a, b)^{(e,f)}$ and requires no further attention. The second case allows the computation of $c = e - wf + wd$. Applying this result to $d + wc$ yields $d + wc = d + w(e - wf + wd) = d(1 + w^2) - w^2 f + we$. To satisfy $(a, b)^{(c,d)} = (a, b)^{(e,f)}$, the equation $d + wc = f + we$ must hold. Setting $d + wc := d(1 + w^2) - w^2 f + we$ in $d + wc = f + we$ the following can be observed:

$$d(1 + w^2) - w^2 f + we = f + we \Leftrightarrow d(1 + w^2) = f(1 + w^2)$$

From the proof of Theorem 4.3.1 it follows that $1 + w^2 \neq 0$ holds as $q \equiv 3 \pmod{4}$, so that $d = f$ must hold for all $w \in \mathbb{Z}_q$ which, however, contradicts the assumption $d \neq f$. The above proof can be formulated for the case that $a \in \mathbb{G}_q$ and $b \in \mathbb{G}_q \setminus \{1\}$ with analogous concluding steps. It is easy to see that $(1, 1)$ is the only element that cannot be used to generate $\mathbb{G}_p$. $\square$

4.3.5   Security Reductions

In this section, the Discrete Logarithm Problem, the Diffie-Hellman Problem and the Decision Diffie-Hellman Problem are defined in terms of fusion-exponentiation. Since the exponents are pairs of integers, each problem is defined twice in terms of the directly available information about the exponent:

1. The *full* exponent (i.e. both components) is kept secret.

2. One *half* of the exponent (i.e. one component) is kept secret.

Figure 4.1 illustrates the relations shown in this section. The proofs are designed from a complexity theory point of view, i.e. the oracles defined in Section 4.3.5 and further oracles defined in this section, are used to perform black-box reductions.

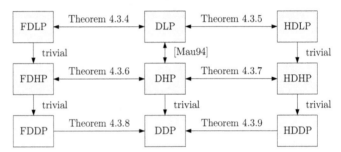

Fig. 4.1: Relations concerning the Fusion Discrete Logarithm Problem Family.

Henceforth, let q be a prime, $q \equiv 3 \pmod 4$, $g \in \mathbb{G}_q \setminus \{1\}$, $\mathbf{g} = (g_1, g_2)$ a generating basis for $\mathbb{G}_p$ and $w = dlog_{g_1}(g_2)$. Notice that $(g, 1)$ is also a generating basis for $\mathbb{G}_p$.

4.3.5.1   The Fusion Discrete Logarithm Problem

In the following, two variants of the Fusion Discrete Logarithm Problem and the corresponding oracles are defined in terms of fusion-exponentiation.

Definition 4.3.1 Let $\mathsf{y} = \mathbf{g}^{\mathsf{x}}$, where $\mathsf{x} \in \mathbb{F}_p$ and $\mathsf{x} = (x_1, x_2)$. Moreover, let $i \in \{1, 2\}$. The Discrete Logarithm Problem in $\mathbb{G}_p$ is twofold:

1. The *Full Fusion Discrete Logarithm Problem* (FDLP) is the following: given y, $\mathbf{g}$ and q, find x. The corresponding oracle is $\mathcal{O}_{\mathrm{FDLP}}(\mathsf{y}, \mathbf{g}) = \mathsf{x}$.

2. The *Half Fusion Discrete Logarithm Problem* (HDLP) is the following: given y, x_i, i, $\mathbf{g}$ and q, find x_{3-i}. The corresponding oracle is $\mathcal{O}_{\mathrm{HDLP}}(\mathsf{y}, x_i, i, \mathbf{g}) = x_{3-i}$.

The following two theorems show that both variants are computationally equivalent to the ordinary Discrete Logarithm Problem in $\mathbb{G}_q$.

Theorem 4.3.4 FDLP $\equiv_p$ DLP

Proof. It has to be shown that (1) FDLP $\leq_p$ DLP and (2) DLP $\leq_p$ FDLP hold.

(1) Let $(y_1, y_2) = (g_1, g_2)^{(x_1, x_2)} = (g_1^{x_1} g_2^{-x_2}, g_1^{x_2} g_2^{x_1})$, where $(x_1, x_2) \in \mathbb{F}_p$. Querying $\mathcal{O}_{DLP}(g_2, g_1)$, $\mathcal{O}_{DLP}(y_1, g_1)$ and $\mathcal{O}_{DLP}(y_2, g_1)$ results in w, $v_1 = x_1 - wx_2$ and $v_2 = x_2 + wx_1$ respectively. From the two equations one can compute $x_1 = (v_1 + wv_2)(1 + w^2)^{-1}$ and $x_2 = v_2 - wx_1$, since $-1 \notin QR_q$.

(2) Let $y = g^x$, where $x \in \mathbb{Z}_q$. Querying $\mathcal{O}_{FDLP}((y, 1), (g, 1))$ results in $(x, 0)$, since $(g, 1)^{(x, 0)} = (g^x 1^{-0}, g^0 1^x) = (g^x, g^0) = (g^x, 1) = (y, 1)$. $\square$

Theorem 4.3.5 HDLP $\equiv_p$ DLP

Proof. It has to be shown that (1) HDLP $\leq_p$ DLP and (2) DLP $\leq_p$ HDLP hold.

(1) Let $(y_1, y_2) = (g_1, g_2)^{(x_1, x_2)} = (g_1^{x_1} g_2^{-x_2}, g_1^{x_2} g_2^{x_1})$, where $(x_1, x_2) \in \mathbb{F}_p$ and x_i is known for any (fixed) $i \in \{1, 2\}$. Furthermore, let $v_1 = x_1 - wx_2$ and $v_2 = x_2 + wx_1$. Querying $\mathcal{O}_{DLP}(g_2, g_1)$ and $\mathcal{O}_{DLP}(y_{3-i}, g_1)$ gives w and v_{3-i}. We set $u_i := (-1)^i$ (computed in $\mathbb{Z}$), so that computing $v_{3-i} + u_i wx_i$ results in x_{3-i}:

$$v_{3-i} + u_i wx_i = \begin{cases} v_2 - wx_1 = x_2 + wx_1 - wx_1 = x_2 & \text{if } i = 1 \\ v_1 + wx_2 = x_1 - wx_2 + wx_2 = x_1 & \text{if } i = 2 \end{cases}$$

(2) Let $y = g^x$, where $x \in \mathbb{Z}_q$. Querying $\mathcal{O}_{HDLP}((y, 1), 0, 2, (g, 1))$ results in x, since $(g, 1)^{(x, 0)} = (y, 1)$ (cf. (2) in the proof of Theorem 4.3.4). $\square$

4.3.5.2 The Fusion Diffie-Hellman Problem

The Diffie-Hellman Problem can also be formulated in two ways in terms of fusion-exponentiation. Again the definitions include the corresponding oracles.

Definition 4.3.2 Let $y_i = g^{x_i}$, where $x_i \in \mathbb{F}_p$ and $x_i = (x_{i1}, x_{i2})$, for $i = 1, 2$. Moreover, let $j, k \in \{1, 2\}$. The Diffie-Hellman Problem in $\mathbb{G}_p$ is twofold:

1. The *Full Fusion Diffie-Hellman Problem* (FDHP) is the following: given y_1, y_2, g and q, find $y_3 = g^{x_1 x_2}$. The corresponding oracle is $\mathcal{O}_{FDHP}(y_1, y_2, g) = g^{x_1 x_2}$.

2. The *Half Fusion Diffie-Hellman Problem* (HDHP) is the following: given y_1, y_2, x_{1j}, x_{2k}, j, k, g and q, find $y_3 = g^{x_1 x_2}$. The corresponding oracle is the following:

$$\mathcal{O}_{HDHP}(y_1, y_2, x_{1j}, x_{2k}, j, k, g) = g^{x_1 x_2}$$

Again it can be shown that the two defined problems are equivalent to their pendant in $\mathbb{G}_q$. This time, however, the reductions are more complex. To simplify several steps, we assert the following: computing $\mathsf{x}_1\mathsf{x}_2 = (x_{11}, x_{12})(x_{21}, x_{22})$ gives the pair $(x_{11}x_{21} - x_{12}x_{22}, x_{11}x_{22} + x_{12}x_{21})$, so that $\mathsf{y}_3 = \mathsf{g}^{\mathsf{x}_1\mathsf{x}_2}$ gives the following:

$$(y_{31}, y_{32}) = (g_1^{x_{11}x_{21} - x_{12}x_{22}} g_2^{-x_{11}x_{22} - x_{12}x_{21}}, g_1^{x_{11}x_{22} + x_{12}x_{21}} g_2^{x_{11}x_{21} - x_{12}x_{22}}) \qquad (4.3.12)$$

At the relevant steps, Equation (4.3.12) is used to simplify the stated conclusions.

Theorem 4.3.6 FDHP $\equiv_p$ DHP

Proof. It has to be shown that (1) FDHP $\leq_p$ DHP and (2) DHP $\leq_p$ FDHP hold.

(1) Let $(y_{i1}, y_{i2}) = (g_1, g_2)^{(x_{i1}, x_{i2})} = (g_1^{x_{i1}} g_2^{-x_{i2}}, g_1^{x_{i2}} g_2^{x_{i1}}) = (g_1^{x_{i1} - wx_{i2}}, g_1^{x_{i2} + wx_{i1}})$, where $(x_{i1}, x_{i2}) \in \mathbb{F}_p$, for $i = 1, 2$. Now, one can compute:

$$A = \mathcal{O}_{\text{DHP}}(y_{11}, g_2, g_1) = \mathcal{O}_{\text{DHP}}(g_1^{x_{11} - wx_{12}}, g_1^w, g_1) = g_1^{(x_{11} - wx_{12})w} = g_1^{x_{11}w - w^2 x_{12}}$$

$$B = \mathcal{O}_{\text{DHP}}(y_{12}, g_2, g_1) = \mathcal{O}_{\text{DHP}}(g_1^{x_{12} + wx_{11}}, g_1^w, g_1) = g_1^{(x_{12} + wx_{11})w} = g_1^{x_{12}w + w^2 x_{11}}$$

$$C = \mathcal{O}_{\text{DHP}}(y_{21}, g_2, g_1) = \mathcal{O}_{\text{DHP}}(g_1^{x_{21} - wx_{22}}, g_1^w, g_1) = g_1^{(x_{21} - wx_{22})w} = g_1^{x_{21}w - w^2 x_{22}}$$

$$D = \mathcal{O}_{\text{DHP}}(y_{22}, g_2, g_1) = \mathcal{O}_{\text{DHP}}(g_1^{x_{22} + wx_{21}}, g_1^w, g_1) = g_1^{(x_{22} + wx_{21})w} = g_1^{x_{22}w + w^2 x_{21}}$$

For the following we need to obtain $W = \mathcal{O}_{\text{DHP}}(g_2, g_2, g_1) = g_1^{w^2}$. Let $\widetilde{W} = W g_1 = g_1^{w^2 + 1}$. From Proposition 4.2.5, Div-DHP $\equiv_p$ DHP, and so one can compute:

$$\widetilde{A} = \mathcal{O}_{\text{Div-DHP}}(Ay_{12}^{-1}, \widetilde{W}^{-1}, g_1) = \mathcal{O}_{\text{Div-DHP}}(g_1^{x_{11}w - w^2 x_{12}} g_1^{-(x_{12} + wx_{11})}, g_1^{-(w^2 + 1)}, g_1)$$

$$= \mathcal{O}_{\text{Div-DHP}}(g_1^{x_{12}(-w^2 - 1)}, g_1^{-w^2 - 1}, g_1) = g_1^{x_{12}(-w^2 - 1)(-w^2 - 1)^{-1}} = g_1^{x_{12}}$$

$$\widetilde{B} = \mathcal{O}_{\text{Div-DHP}}(By_{11}, \widetilde{W}, g_1) = \mathcal{O}_{\text{Div-DHP}}(g_1^{x_{12}w + w^2 x_{11}} g_1^{x_{11} - wx_{12}}, g_1^{w^2 + 1}, g_1)$$

$$= \mathcal{O}_{\text{Div-DHP}}(g_1^{x_{11}(w^2 + 1)}, g_1^{w^2 + 1}, g_1) = g_1^{x_{11}(w^2 + 1)(w^2 + 1)^{-1}} = g_1^{x_{11}}$$

$$\widetilde{C} = \mathcal{O}_{\text{Div-DHP}}(Cy_{22}^{-1}, \widetilde{W}^{-1}, g_1) = \mathcal{O}_{\text{Div-DHP}}(g_1^{x_{21}w - w^2 x_{22}} g_1^{-(x_{22} + wx_{21})}, g_1^{-(w^2 + 1)}, g_1)$$

$$= \mathcal{O}_{\text{Div-DHP}}(g_1^{x_{22}(-w^2 - 1)}, g_1^{-w^2 - 1}, g_1) = g_1^{x_{22}(-w^2 - 1)(-w^2 - 1)^{-1}} = g_1^{x_{22}}$$

$$\widetilde{D} = \mathcal{O}_{\text{Div-DHP}}(Dy_{21}, \widetilde{W}, g_1) = \mathcal{O}_{\text{Div-DHP}}(g_1^{x_{22}w + w^2 x_{21}} g_1^{x_{21} - wx_{22}}, g_1^{w^2 + 1}, g_1)$$

$$= \mathcal{O}_{\text{Div-DHP}}(g_1^{x_{21}(w^2 + 1)}, g_1^{w^2 + 1}, g_1) = g_1^{x_{21}(w^2 + 1)(w^2 + 1)^{-1}} = g_1^{x_{21}}$$

These intermediate results enable the following computations:

$$E = \mathcal{O}_{\text{DHP}}(\widetilde{B}, \widetilde{D}, g_1) = \mathcal{O}_{\text{DHP}}(g_1^{x_{11}}, g_1^{x_{21}}, g_1) = g_1^{x_{11}x_{21}}$$

$$F = \mathcal{O}_{\text{DHP}}(\widetilde{A}, \widetilde{C}, g_1) = \mathcal{O}_{\text{DHP}}(g_1^{x_{12}}, g_1^{x_{22}}, g_1) = g_1^{x_{12}x_{22}}$$

$$G = \mathcal{O}_{\text{DHP}}(\widetilde{B}, \widetilde{C}, g_1) = \mathcal{O}_{\text{DHP}}(g_1^{x_{11}}, g_1^{x_{22}}, g_1) = g_1^{x_{11}x_{22}}$$

$$H = \mathcal{O}_{\text{DHP}}(\widetilde{A}, \widetilde{D}, g_1) = \mathcal{O}_{\text{DHP}}(g_1^{x_{12}}, g_1^{x_{21}}, g_1) = g_1^{x_{12}x_{21}}$$

Finally, y_{31} and y_{32}, as given in Equation (4.3.12), can be computed as follows:

$$y_{31} = EF^{-1}(\mathcal{O}_{\mathrm{DHP}}(G, g_2, g_1)\mathcal{O}_{\mathrm{DHP}}(H, g_2, g_1))^{-1}$$
$$= g_1^{x_{11}x_{21}} g_1^{-x_{12}x_{22}}(\mathcal{O}_{\mathrm{DHP}}(g_1^{x_{11}x_{22}}, g_1^w, g_1)\mathcal{O}_{\mathrm{DHP}}(g_1^{x_{12}x_{21}}, g_1^w, g_1))^{-1}$$
$$= g_1^{x_{11}x_{21} - x_{12}x_{22}}(g_1^{x_{11}x_{22}w} g_1^{x_{12}x_{21}w})^{-1}$$
$$= g_1^{x_{11}x_{21} - x_{12}x_{22}}(g_1^w)^{-x_{11}x_{22} - x_{12}x_{21}}$$
$$= g_1^{x_{11}x_{21} - x_{12}x_{22}} g_2^{-x_{11}x_{22} - x_{12}x_{21}}$$
$$y_{32} = GH\mathcal{O}_{\mathrm{DHP}}(E, g_2, g_1)\mathcal{O}_{\mathrm{DHP}}(F, g_2, g_1)^{-1}$$
$$= g_1^{x_{11}x_{22}} g_1^{x_{12}x_{21}} \mathcal{O}_{\mathrm{DHP}}(g_1^{x_{11}x_{21}}, g_1^w, g_1)\mathcal{O}_{\mathrm{DHP}}(g_1^{x_{12}x_{22}}, g_1^w, g_1)^{-1}$$
$$= g_1^{x_{11}x_{22} + x_{12}x_{21}} g_1^{x_{11}x_{21}w} g_1^{-x_{12}x_{22}w}$$
$$= g_1^{x_{11}x_{22} + x_{12}x_{21}}(g_1^w)^{x_{11}x_{21} - x_{12}x_{22}}$$
$$= g_1^{x_{11}x_{22} + x_{12}x_{21}} g_2^{x_{11}x_{21} - x_{12}x_{22}}$$

(2) Let $y_i = g^{x_i}$, where $x_i \in \mathbb{Z}_q$, for $i = 1, 2$. Querying $\mathcal{O}_{\mathrm{FDHP}}((y_1, 1), (y_2, 1), (g, 1))$ results in $y_3 = (g^{x_1 x_2}, 1)$, since $(y_1, 1) = (g, 1)^{(x_1, 0)}$, $(y_2, 1) = (g, 1)^{(x_2, 0)}$ and $(x_1, 0)(x_2, 0) = (x_1 x_2, 0)$. $\qquad\square$

Theorem 4.3.7 HDHP $\equiv_p$ DHP

Proof. It has to be shown that (1) HDHP $\leq_p$ DHP and (2) DHP $\leq_p$ HDHP hold.

(1) Let $(y_{i1}, y_{i2}) = (g_1, g_2)^{(x_{i1}, x_{i2})} = (g_1^{x_{i1}} g_2^{-x_{i2}}, g_1^{x_{i2}} g_2^{x_{i1}}) = (g_1^{x_{i1} - w x_{i2}}, g_1^{x_{i2} + w x_{i1}})$, where $(x_{i1}, x_{i2}) \in \mathbb{F}_p$, for $i = 1, 2$. Without loosing generality we consider the case where x_{11} and x_{21} are known, i.e. $j = k = 1$. Then, one can compute

$$A = y_{11} g_1^{-x_{11}} = (g_1^{x_{11}} g_2^{-x_{12}}) g_1^{-x_{11}} = g_2^{-x_{12}} \qquad B = y_{12} g_2^{-x_{11}} = (g_1^{x_{12}} g_2^{x_{11}}) g_2^{-x_{11}} = g_1^{x_{12}}$$
$$C = y_{21} g_1^{-x_{21}} = (g_1^{x_{21}} g_2^{-x_{22}}) g_1^{-x_{21}} = g_2^{-x_{22}} \qquad D = y_{22} g_2^{-x_{21}} = (g_1^{x_{22}} g_2^{x_{21}}) g_2^{-x_{21}} = g_1^{x_{22}}$$

These intermediate results enable the following computations:

$$E = \mathcal{O}_{\mathrm{DHP}}(B, D, g_1) = \mathcal{O}_{\mathrm{DHP}}(g_1^{x_{12}}, g_1^{x_{22}}, g_1) = g_1^{x_{12}x_{22}}$$
$$F = \mathcal{O}_{\mathrm{DHP}}(A, C, g_2) = \mathcal{O}_{\mathrm{DHP}}(g_2^{-x_{12}}, g_2^{-x_{22}}, g_2) = g_2^{x_{12}x_{22}}$$

Finally, y_{31} and y_{32}, as given in Equation (4.3.12), can be computed as follows:

$$y_{31} = g_1^{x_{11}x_{21}} E^{-1} C^{x_{11}} A^{x_{21}} = g_1^{x_{11}x_{21}} g_1^{-x_{12}x_{22}}(g_2^{-x_{22}})^{x_{11}}(g_2^{-x_{12}})^{x_{21}}$$
$$= g_1^{x_{11}x_{21} - x_{12}x_{22}} g_2^{-x_{11}x_{22} - x_{12}x_{21}}$$
$$y_{32} = D^{x_{11}} B^{x_{21}} g_2^{x_{11}x_{21}} F^{-1} = (g_1^{x_{22}})^{x_{11}}(g_1^{x_{12}})^{x_{21}} g_2^{x_{11}x_{21}} g_2^{-x_{12}x_{22}}$$
$$= g_1^{x_{11}x_{22} + x_{12}x_{21}} g_2^{x_{11}x_{21} - x_{12}x_{22}}$$

If $j = k = 2$, or $j = 1$ $(j = 2)$ and $k = 2$ $(k = 1)$, the reductions work similarly.

(2) Let $y_i = g^{x_i}$, where $x_i \in \mathbb{Z}_q$, for $i = 1, 2$. Querying $\mathcal{O}_{\mathrm{HDHP}}((y_1, 1), (y_2, 1), 0, 0, 2, 2, (g, 1))$ we get $(y_3, 1)$, where $y_3 = g^{x_1 x_2}$ holds for obvious reasons. $\qquad\square$

4.3.5.3   The Fusion Decision Diffie-Hellman Problem

Once again we can give two definitions for $\mathbb{G}_p$.

Definition 4.3.3 Let $y_i = g^{x_i}$, where $x_i \in \mathbb{F}_p$ and $x_i = (x_{i1}, x_{i2})$, for $i = 1, 2, 3$. Moreover, let $j, k, l \in \{1, 2\}$. The Decision Diffie-Hellman Problem in $\mathbb{G}_p$ is twofold:

1. The *Full Fusion Decision Diffie-Hellman Problem* (FDDP) is the following: given y_1, y_2, y_3, g and q, decide if $x_3 = x_1 x_2$ holds. The corresponding oracle is $\mathcal{O}_{\text{FDDP}}(y_1, y_2, y_3, g) = b$, where $b = 1$ if $x_3 = x_1 x_2$ and $b = 0$ otherwise.

2. The *Half Fusion Decision Diffie-Hellman Problem* (HDDP) is the following: given y_1, y_2, y_3, x_{1j}, x_{2k}, x_{3l}, j, k, l, g and q, decide if $x_3 = x_1 x_2$. The corresponding oracle is $\mathcal{O}_{\text{HDDP}}(y_1, y_2, y_3, x_{1j}, x_{2k}, x_{3l}, j, k, l, g) = b$, where $b = 1$ if $x_3 = x_1 x_2$ and $b = 0$ otherwise.

Theorem 4.3.8 DDP $\leq_p$ FDDP

Proof. Let $y_i = g^{x_i}$, where $x_i \in \mathbb{Z}_q$, for $i = 1, 2, 3$. Querying $\mathcal{O}_{\text{FDDP}}((y_1, 1), (y_2, 1), (y_3, 1), (g, 1))$ results in 1, if $(x_3, 0) = (x_1, 0)(x_2, 0)$, and 0 otherwise.   □

Theorem 4.3.9 DDP $\leq_p$ HDDP

Proof. Let $y_i = g^{x_i}$, where $x_i \in \mathbb{Z}_q$, for $i = 1, 2, 3$. Querying $\mathcal{O}_{\text{HDDP}}((y_1, 1), (y_2, 1), (y_3, 1), 0, 0, 0, 2, 2, 2, (g, 1))$ results in 1, if $(x_3, 0) = (x_1, 0)(x_2, 0)$ and 0 otherwise.   □

4.3.5.4   Interpretation

If the security of a cryptosystem relies on the DLP or the DHP in $\mathbb{G}_q$, then it can be based on $\mathbb{G}_p$ while maintaining the same computational assumptions. This directly follows from DLP $\equiv_p$ FDLP $\equiv_p$ HDLP and DHP $\equiv_p$ FDHP $\equiv_p$ HDHP. Interestingly, even if one half of a fusion-exponent is known, achieving the desired attack goals is at least as hard as breaking the DLP or DHP in $\mathbb{G}_q$.

More attention is required if a scheme is formerly based on the Decision Diffie-Hellman Problem in $\mathbb{G}_q$. In this problem, a random triple $(y_1, y_2, y_3) \in \mathbb{G}_q$ needs to be indistinguishable from a triple of the form $(g^{x_1}, g^{x_2}, g^{x_1 x_2})$, where $x_1, x_2 \in \mathbb{Z}_q$. The computational equivalences between the FDDP, HDDP and the DDP have neither been proven nor disproven and remain as an open problem. However, if the FDDP or the HDDP can be solved, then the DDP can be solved as well. So, the DDP seems to be a stronger assumption than its variants in $\mathbb{G}_p$.

For the Discrete Logarithm Problem in $\mathbb{G}_p$ one can define a third variant where y and g are given, and the goal is to find either x_1 or x_2, such that $y = g^{(x_1, x_2)}$ holds. In Section 5.6, a protocol is given which is proven to be secure under the assumption that this third variant is also hard to solve. In Lemma 5.6.1, this problem is proven to be computationally equivalent to the Discrete Logarithm Problem in $\mathbb{G}_q$.

Some trivial reductions that are not given here for simplicity are HDDP $\leq_p$ FDDP, FDHP $\leq_p$ FDLP, HDHP $\leq_p$ HDLP, FDDP $\leq_p$ FDHP and HDDP $\leq_p$ HDHP.

4.4   Fusion based on a Group of Hidden Order

Applications exist where discrete-logarithm-based and RSA-like cryptosystems are used in the same algebraic setting. For instance, one can verifiably encrypt RSA-signatures using ElGamal encryption if run over the same composite modulus [CG98]. Furthermore, interactive interval-proofs can be improved regarding efficiency [Bou00]. Such proofs are a building block in several cryptosystems, such as given in [ACJT00]. In such schemes, it is important that the requirements of both paradigms are preserved:

1. To keep solving the Discrete Logarithm Problem, the Diffie-Hellman Problem and the Decision Diffie-Hellman Problem hard, an intrinsic requirement is that the order of the used group is large and has at least one large prime factor.

2. For RSA-like schemes, it is important that finding the order of the group is hard.

As mentioned in Section 4.2.2, it is advantageous to use QR_n in this context, where n is the product of two large safe-primes $p = 2p' + 1$ and $q = 2q' + 1$ [Bon98, BBR99]. Remember that the order of QR_n is $\tilde{n} = p'q'$.

4.4.1   Exponents that are Pairs of Integers

In Theorem 4.3.1, we proved that $\mathbb{Z}_q[X]/(X^2 + 1)$ is a field for q prime and $q \equiv 3 \pmod 4$. For $\mathbb{Z}_{\tilde{n}}[X]/(X^2 + 1)$ we get a weaker – but for most discrete-logarithm-based cryptosystems sufficient – result:

Theorem 4.4.1 $\mathbb{Z}_{\tilde{n}}[X]/(X^2 + 1)$ *is a commutative unitary ring.*

Proof. For $\tilde{n}$ being a natural number, $\mathbb{Z}_{\tilde{n}}$ is a commutative unitary ring and thus so $\mathbb{Z}_{\tilde{n}}[X]$. Together with $(X^2 + 1)$ being an ideal $\mathbb{Z}_{\tilde{n}}[X]/(X^2 + 1)$ is a commutative unitary ring.   □

Henceforth, we set $\hat{n} = \tilde{n}^2$ and $R_{\hat{n}} := \mathbb{Z}_{\tilde{n}}[X]/(X^2 + 1)$. Notice that computations in $R_{\hat{n}}$ can only be performed if $\tilde{n}$ is known, which is generally *not* the case for discrete-logarithm-based schemes running over QR_n. At least some instances in such a scheme are not allowed to know the order, since otherwise the scheme does not need to be based on QR_n. For the instances who need to know the order, it might be necessary to invert elements in $R_{\hat{n}}$. One such example would be RSA. Then, it must be verified for a chosen pair $(a, b) \in R_{\widehat{n}}$, if $(a^2 + b^2)$ is co-prime to $\widetilde{n}$. This comes from the fact that the multiplicative inverse of (a, b) in $R_{\widehat{n}}$ is

$$(a, b)^{-1} = (a(a^2 + b^2)^{-1}, -b(a^2 + b^2)^{-1}).$$

4.4.2 Fusion-Exponentiation

The fusion exponentiation function ξ can be defined analogously to Section 4.3.3. $\mathbb{G}_{\widehat{n}} := QR_n^2$ with component-wise multiplication forms an abelian group and the properties (4.3.9), (4.3.10) and (4.3.11) can be shown for

$$\xi : \mathbb{G}_{\widehat{n}} \times R_{\widehat{n}} \to \mathbb{G}_{\widehat{n}}, \quad \xi(\mathsf{a}, \mathsf{n}) := (a_1^{n_1} a_2^{-n_2}, a_1^{n_2} a_2^{n_1})$$

since they do not require fusion-exponents to be invertible in $R_{\widehat{n}}$.

4.4.3 Generating Bases

If the order of QR_n is a Blum integer, then a result analogous to that of Theorem 4.3.3 can be achieved: elements of the form (a, b), $(a, 1)$ and $(1, b)$, where $\langle a \rangle = \langle b \rangle = QR_n$, can be used to generate $\mathbb{G}_{\widehat{n}}$ by using fusion-exponentiation. The requirement that the order is a Blum integer is necessary such that a proof analogous to that in Theorem 4.3.3 can be given. The following two results are well known [MVO96]:

Proposition 4.4.1 *Let $n = pq$, where p and q are distinct primes. A natural number a is a quadratic residue modulo n, iff $a \in QR_p$ and $a \in QR_q$.*

Proposition 4.4.2 *Let a and b be two generators of QR_n. Then $\gcd(dlog_a(b), \widetilde{n}) = 1$.*

Based on the above two propositions and Theorem 4.3.3, one can state the Theorem:

Theorem 4.4.2 $\mathbb{G}_{\widehat{n}} = \{(a, b)^{(c,d)} \mid (a, b) \in \{1, g \mid \langle g \rangle = QR_n\}^2 \setminus \{(1, 1)\}, (c, d) \in R_{\widehat{n}}\}$.

Proof. Essentially, the proof works analogously to that of Theorem 4.3.3. The following two conclusions in the proof of Theorem 4.3.3 require special attention, since now the order of the underlying group QR_n is $\widetilde{n}$ and hence *not prime*:

1. From $c = e$ it has been followed that $d = f$ needs to hold for $(a, b)^{(c,d)} = (a, b)^{(e,f)}$. This was possible, because $w = dlog_a(b)$ was co-prime to the order q.

2. From $d(1 + w^2) = f(1 + w^2)$ it has been followed that $d = f$. This was only possible, because $1 + w^2$ was co-prime to q. The latter held by $q \equiv 3 \pmod 4$.

Now consider these two conclusions for the current situation where the order is $\widetilde{n} = p'q'$ and a Blum integer. From Proposition 4.4.2 it follows that w is co-prime to $\widetilde{n}$ and hence the first conclusion holds. For the second conclusion it must be shown that

$gcd(1 + w^2, \tilde{n}) = 1$. Firstly assume that $q'|(1 + w^2)$, i.e. an integer k exists such that $1 + w^2 = kq'$. Then, $w^2 = kq' - 1$ and so $kq' - 1 \in QR_{\tilde{n}}$ must hold. Proposition 4.4.1 implies that $kq' - 1 \in QR_{q'}$ and hence $-1 \in QR_{q'}$. From Corollary 4.2.1, however, this gives a contradiction, since $\tilde{n}$ is a Blum integer, i.e. $q' \equiv 3 \pmod 4$. The same can be shown for $p'|(1 + w^2)$. Together, this gives $gcd(1 + w^2, \tilde{n}) = 1$ and conclusion two holds. The remainder works analogously to Theorem 4.3.3. □

4.4.4 Security Reductions

The security reductions of Section 4.3.5 also hold for $\mathbb{G}_{\hat{n}}$ if $\tilde{n}$ is a Blum integer. In the proofs of the Theorems 4.3.4 and 4.3.6, $1 + w^2$ is required to be co-prime to the group order. In the proof of Theorem 4.4.2, this has been shown to hold if $\tilde{n}$ is a Blum integer. Furthermore, it is important that $w = dlog_{g_1}(g_2)$ is co-prime to $\tilde{n}$. This is also covered by Theorem 4.4.2. All other steps performed in the reductions require only the algebraic properties that have been shown for $R_{\hat{n}}$ and $\mathbb{G}_{\hat{n}}$.

4.5 Efficiency

In this section, fusion-exponentiation and ordinary exponentiation are compared in terms of efficiency. It is shown that in special cases, fusion-exponentiation is more efficient than ordinary exponentiation. Furthermore, the question of whether a simplified generating basis can speed-up computations is discussed. Finally, an optimization for fusion-multiplication is considered. The efficiency estimations deal with concrete bit-lengths and numbers of operations, which can be considered as an average case.

4.5.1 Ordinary versus Fusion-Exponentiation

One fusion-exponentiation requires four ordinary exponentiations. Hence, it is obviously more expensive if run over the *same* system parameters. In this section, efficiency estimations are done for the following three types of groups:

1. $\mathbb{G}_q$ as a cyclic subgroup of $\mathbb{Z}_{p'}^*$, where q and p' are primes, such that $q|(p' - 1)$.

2. $\mathbb{G}_q$ as a cyclic subgroup of the elliptic curve group $E(\mathbb{Z}_{p'})$, where $q|\#E(\mathbb{Z}_{p'})$.

3. QR_n, where n is the product of two safe-primes.

Let l_{q_0}, $l_{p'_0}$ and l_{n_0} denote the lower security bounds for q, p' and n in terms of their bit-lengths. As stated in Table 3.4, the recommended bounds are $l_{q_0} = 160$ and $l_{p'_0} = 1024$ for $\mathbb{G}_q < \mathbb{Z}_{p'}^*$, $l_{q_0} = 180$ and $l_{p'_0} = 192$ for $\mathbb{G}_q < E(\mathbb{Z}_{p'})$, and $l_{n_0} = 1024$ for QR_n. In

[PT05], no direct suggestions are given for QR_n, but rather for $\mathbb{Z}_n$ with the same type of modulus. Since QR_n is used for discrete-logarithm-based cryptosystems in the same setting as RSA-like cryptosystems, the same lower bound can be used.

To avoid confusion when comparing ordinary with fusion-exponentiation, the parameters q, p' and n for the different settings are distinguished as follows: the parameters q_1, p'_1 and n_1 are used for the ordinary settings $\mathbb{G}_{q_1}$ and QR_{n_1}, whereas q_2, p'_2 and n_2 are used for the fusion-settings $\mathbb{G}_{p_2}$ and $\mathbb{G}_{\widehat{n}_2}$, where $p_2 = q_2^2$ and $\widehat{n}_2 = \widetilde{n}_2^2$.

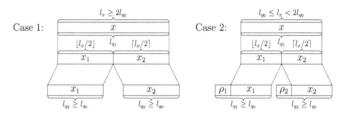

Fig. 4.2: Ordinary Setting $\mathbb{G}_{q_1}$ versus Fusion-Setting $\mathbb{G}_{p_2}$.

In the following, an abstract description is given describing how the ordinary setting and the fusion-setting are compared. Here, $\mathbb{G}_q$ is considered for simplicity (discussions for QR_n are analogous). Let x be a number, that lies in the interval $[0, 2^{l_x} - 1]$, where $l_x \geq l_{q0}$. The value x may be expanded using padding up to a required bit-length, but it is strictly prohibited to shorten x as this results in important information being lost. For instance, x is the output of a collision-free number generator. Now the bits of x are taken as the exponent on the one hand in $\mathbb{G}_{q_1}$ and on the other hand in $\mathbb{G}_{p_2}$.

1. *Ordinary Exponentiation in $\mathbb{G}_{q_1}$:* The prime q_1 needs to be chosen such that $q_1 > x$ always holds, otherwise x might be destroyed through a modulo reduction. For simplicity, we set $l_x = l_{q_1}$ and perform a check to ensure that $x < q_1$. In this setting, only one exponentiation in $\mathbb{G}_{q_1}$ is necessary:

$$y = g^x$$

2. *Fusion-Exponentiation in $\mathbb{G}_{p_2}$:* Here, x is cut into two parts x_1 and x_2 of approximately equal length, i.e. $l_{x_1} = \lfloor l_x/2 \rfloor$ and $l_{x_2} = \lceil l_x/2 \rceil$. If $l_x \geq 2l_{q0}$ (cf. Figure 4.2, Case 1) then q_2 needs to be chosen such that $q_2 > x_1$ and $q_2 > x_2$ always hold, otherwise either x_1 or x_2 might be destroyed through a modulo reduction. This gives the following fusion-exponentiation in $\mathbb{G}_{p_2}$:

$$(y_1, y_2) = (g_1, g_2)^{(x_1, x_2)}$$

For the case that $l_x \geq l_{q0}$, but $l_x < 2l_{q0}$ (cf. Figure 4.2, Case 2), then it is necessary to expand x_1 and x_2 up to a minimum of approximately l_{q0} bits. Let $l_{q_2} \geq l_{q0}$, then

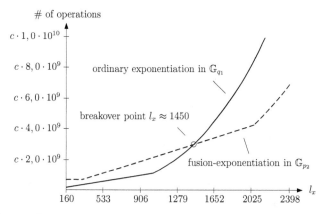

Fig. 4.3: Square-and-Multiply with Standard Multiplication in Cyclic Subgroups of $\mathbb{Z}_{p'}^*$.

we get the following fusion-exponentiation in $\mathbb{G}_{p_2}$:

$$(y_1, y_2) = (g_1, g_2)^{(\rho_1 2^{l_{x_1}} + x_1, \rho_2 2^{l_{x_2}} + x_2)}$$

where ρ_i is a random padding up to l_{q_2} bits, with the condition $\rho_i 2^{l_{x_i}} + x_i < q_2$.

So we have one exponentiation with an l_{q_1}-bit exponent against four exponentiations with l_{q_2}-bit exponents, where $l_{q_1} \leq 2l_{q_2}$. Now the breakover point for l_x needs to be found, where fusion-exponentiation in $\mathbb{G}_{p_2}$ starts to be more efficient than ordinary exponentiation in $\mathbb{G}_{q_1}$. The same needs to be analyzed for QR_{n_1} and $\mathbb{G}_{\hat{n}_2}$.

4.5.1.1 Cyclic Subgroups of $\mathbb{Z}_{p'}^*$

We start by setting $l_x = l_{q_1} = l_{q_2} = l_{q_0} = 160$ and $l_{p_1'} = l_{p_2'} = l_{p_0'} = 1024$. When increasing l_x bit-wise, one has to increase l_{q_1} bit-wise as well, but l_{q_2} remains unchanged until $l_x = 2l_{q_0} = 320$. Then, l_{q_2} is increased every second time l_x is increased. Notice that $l_{p_1'}$ and $l_{p_2'}$ are not changed until l_{q_1} or l_{q_2} exceeds 1024. In Figure 4.3, one can see the number of operations that are necessary for ordinary exponentiation in $\mathbb{G}_{q_1}$ and for fusion-exponentiation in $\mathbb{G}_{p_2}$, if square-and-multiply is used based on standard multiplication (neglecting a constant c here). It can be seen that the breakover point, where fusion-exponentiation begins to be more efficient, is $l_x \approx 1450$. So the difference to the current lower security bound $l_{q_0} \approx 160$ is around a factor of 9. Note that in Figure 4.3, the curve for fusion-exponentiation has two sharp bends, one at $l_x = 320$ and one at $l_x = 2048$. The first occurs because until $l_x = 320$, $l_{q_1} \leq 2l_{q_2}$ holds without changing l_{q_2}. The second happens because when $l_{q_2} = 1024$ is reached, one has to start increasing $l_{p_2'}$. The curve for ordinary exponentiation has only one sharp bend at $l_x = 1024$ where $l_{q_1} = 1024$, where one has to start increasing $l_{p_1'}$.

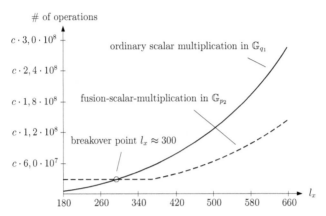

Fig. 4.4: Double-and-Add with Standard Multiplication in Cyclic Subgroups of $E(\mathbb{Z}_{p'})$.

As long as $l_{p'_1} = l_{p'_2}$, a fusion-based encryption scheme can double the throughput as a by-product. Consider fusion-based ElGamal encryption, for instance: in $\mathbb{G}_{q_1}$, the length of a plaintext-block is $l_{p'_1} = 1024$. In $\mathbb{G}_{p_2}$, the length of a plaintext-block is $2l_{p'_2} = 2048$. This yields an additional speed-up if larger plaintexts are encrypted.

4.5.1.2 Cyclic Subgroups of $E(\mathbb{Z}_{p'})$

Compared to cyclic subgroups of $\mathbb{Z}_{p'}^*$, this situation is quite different because l_q and $l_{p'}$ are close to being equal from the beginning. Here, (when $l_{q_1} = 192$ or $l_{q_2} = 192$) scaling l_{q_i} upwards implies increasing $l_{p'_i}$. We start by setting $l_x = l_{q_1} = l_{q_2} = l_{q_0} = 180$ and $l_{p'_1} = l_{p'_2} = l_{p'_0} = 192$. Until $l_x = 360$, neither l_{q_2} nor $l_{p'_2}$ needs to be increased, whereas l_{q_1} is increased from the beginning and $l_{p'_1}$ once $l_{q_1} = 192$ has been reached. Notice that if l_x exceeds 360, then only l_{q_2} needs to be increased by 1 every second time l_{q_1} and $l_{p'_1}$ are increased by 1. However, if l_x exceeds 384, then both l_{q_2} and $l_{p'_2}$ need to be increased by 1 every second time l_{q_1} and $l_{p'_1}$ are increased by 1. In Figure 4.4 it can be seen, that the breakover point, where fusion scalar multiplication begins to be more efficient, is $l_x \approx 300$. Analogous to the previous section, double-and-add (square-and-multiply over points) is used with standard multiplication. This time, l_{q_0} and l_x only differ by a factor of about 1.5. Such a low factor shows the practical relevance of the fusion-algebra.

4.5.1.3 Cyclic Group QR_n

In QR_n, the modulus n is the product of two safe-primes $p = 2p' + 1$ and $q = 2q' + 1$. It has been shown that the order of QR_n is $\tilde{n} = p'q'$. Since the lower security bound for n is $l_{n_0} = 1024$, one obtains $1022 \le l_{\tilde{n}} \le 1024$. Let $l_{\tilde{n}} = 1022$. It is clear that l_{n_1}

needs to be increased from the beginning on, since we start at $l_{n_1} = l_{n_0} = 1024$. Until $l_x = 2044$, l_{n_2} does not need to be increased, since $l_{\bar{n}} = 1022$. The breakover point can be found in Figure 4.5, and is $l_x \approx 1624$. So the difference to the lower security bound is a factor of 1.5, which is a similar result to when elliptic curves are used.

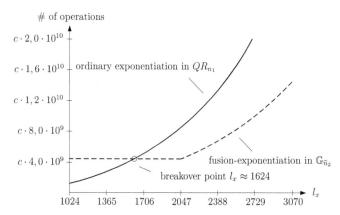

Fig. 4.5: Square-and-Multiply with Standard Multiplication in QR_n.

4.5.1.4   Using More Efficient Multiplication Algorithms

Square-and-multiply can be used with any multiplication algorithm. In the previous sections, standard multiplication has been considered, whose complexity is $O(n^2)$ for n-bit numbers. If Karatsuba's method is used instead of standard multiplication, then the complexity is $O(n^{log_2(3)}) \approx O(n^{1,58})$. If this method is applied for scalar multiplications in cyclic subgroups of $E(\mathbb{Z}_{p'})$, then the new breakover point is around 321 bits. If fast fourier transformations are used, where the complexity is $O(n \log \log n)$, then the breakover point is around 364 bits. Higher breakover points rarely exists.

4.5.2   Using a Simplified Generating Basis

In the previous section, an extreme case was discussed where the generating basis is (g_1, g_2). If the simplified generating basis $(g, 1)$ is used instead, then fusion-exponentiation only requires two ordinary exponentiations, i.e. $(g, 1)^{(a,b)} = (g^a, g^b)$. Notice, however, that then not all components are uniformly fused into results, but if further exponentiations take place, then again four ordinary exponentiations are necessary (and fully fuse the two components into the result). Nevertheless, using $(g, 1)$ can be sufficient to include both key components of a pair uniformly into account *during encryptions*. This can be seen from the following considerations.

ElGamal Encryption with Generating Basis (g_1, g_2): Let $(z_1, z_2) \in \mathbb{F}_p$ be the secret key of the ElGamal encryption scheme, run over $\mathbb{G}_p$. The corresponding public key is defined as $(h_1, h_2) = (g_1, g_2)^{(z_1, z_2)} = (g_1^{z_1} g_2^{-z_2}, g_1^{z_2} g_2^{z_1})$. A plaintext $(m_1, m_2) \in \mathbb{G}_p$ is encrypted as follows:

$$(u_1, u_2) = (g_1, g_2)^{(r_1, r_2)} = (g_1^{r_1} g_2^{-r_2}, g_1^{r_2} g_2^{r_1})$$

$$(e_1, e_2) = (m_1, m_2)(h_1, h_2)^{(r_1, r_2)}$$

$$= (m_1, m_2)(h_1^{r_1} h_2^{-r_2}, h_1^{r_2} h_2^{r_1})$$

$$= (m_1 h_1^{r_1} h_2^{-r_2}, m_2 h_1^{r_2} h_2^{r_1})$$

where (r_1, r_2) is chosen at random from $\mathbb{F}_p$. Given the ciphertext $((u_1, u_2), (e_1, e_2))$ and (z_1, z_2), the plaintext (m_1, m_2) can be obtained as follows:

$$(e_1, e_2)(u_1, u_2)^{(-z_1, -z_2)} = (m_1 h_1^{r_1} h_2^{-r_2}, m_2 h_1^{r_2} h_2^{r_1})(u_1^{-z_1} u_2^{z_2}, u_1^{-z_2} u_2^{-z_1})$$

Notice that this gives (m_1, m_2) because

$$(u_1^{-z_1} u_2^{z_2}, u_1^{-z_2} u_2^{-z_1}) = ((g_1^{r_1} g_2^{-r_2})^{-z_1}(g_1^{r_2} g_2^{r_1})^{z_2}, (g_1^{r_1} g_2^{-r_2})^{-z_2}(g_1^{r_2} g_2^{r_1})^{-z_1})$$

$$= (g_1^{-r_1 z_1} g_2^{r_2 z_1} g_1^{r_2 z_2} g_2^{r_1 z_2}, g_1^{-r_1 z_2} g_2^{r_2 z_2} g_1^{-r_2 z_1} g_2^{-r_1 z_1})$$

$$= ((g_1^{z_1} g_2^{-z_2})^{-r_1}(g_2^{z_2} g_2^{z_1})^{r_2}, (g_1^{z_1} g_2^{-z_2})^{-r_2}(g_1^{z_2} g_2^{z_1})^{-r_1})$$

$$= (h_1^{-r_1} h_2^{r_2}, h_1^{-r_2} h_2^{-r_1})$$

ElGamal Encryption with Generating Basis $(g, 1)$: Let $(z_1, z_2) \in \mathbb{F}_p$ be the secret key and $(h_1, h_2) = (g, 1)^{(z_1, z_2)} = (g^{z_1}, g^{z_2})$ the corresponding public key. A plaintext $(m_1, m_2) \in \mathbb{G}_p$ is encrypted as follows:

$$(u_1, u_2) = (g, 1)^{(r_1, r_2)} = (g^{r_1}, g^{r_2})$$

where (r_1, r_2) is chosen at random from $\mathbb{F}_p$. The remainder of the encryption and the whole decryption work analogously to the setting where (g_1, g_2) is used as the generating basis, and requires the same number of computations.

In Figure 4.6, it can be seen that ElGamal encryption in $\mathbb{G}_{p_2}$ with generating basis $(g, 1)$ has a lower breakover point than the variant where the generating basis is (g_1, g_2). This comes from the fact that 6 ordinary exponentiations are necessary instead of 8.

ElGamal decryption in $\mathbb{G}_{p_2}$ requires 4 exponentiations, independent of whether the basis is $(g, 1)$ or (g_1, g_2). Hence, the breakover point remains around 300 there.

4.5.3   Optimized Fusion-Multiplication

Multiplication in $\mathbb{F}_p$ is expressed by four products and two sums in $\mathbb{Z}_q$. Using Karatsuba's method, these computations can be optimized such that only three products and five sums are necessary. Let $(a, b), (c, d) \in \mathbb{F}_p$ and $(e, f) = (a, b)(c, d) =$

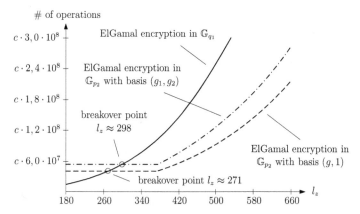

Fig. 4.6: ElGamal Encryption based on Double-and-Add with Standard Multiplication.

$(ac - bd, ad + bc)$. Computing $ad + bc$ can be expressed by $(a + b)(c + d) - ac - bd$. Thus, $\cdot$ only requires three multiplications, which can be seen from the following:

$$x = ac, \quad y = bd, \quad e = x - y, \quad f = (a + b)(c + d) - x - y$$

The correctness of this optimization is obvious and requires no further attention.

4.6 Concluding Remarks

In this chapter, a special algebraic structure has been presented, that can be based on any group $\mathbb{G}_q$ of prime order q, where $q \equiv 3 \pmod 4$. In $\mathbb{G}_p$, a special form of exponentiation has been defined as the *fusion-exponentiation*. Although this kind of exponentiation differs from the ordinary conventions, it follows the same algebraic principles. The problem of inverting the fusion-exponentiation has been defined as the *Fusion Discrete Logarithm Problem*. Beside other reductions, it has been shown that this problem is computationally equivalent to the ordinary Discrete Logarithm Problem in $\mathbb{G}_q$. Furthermore, the *Fusion Diffie-Hellman Problem* has been defined and proven to be computationally equivalent to the ordinary one. Additionally, solving the *Fusion Decision Diffie-Hellman Problem* leads to solving the ordinary one. The converse is currently an open problem.

Beside groups of prime order, the concept of fusion can also be realized for groups of composite order. Finally, it has been shown that fusion-exponentiation is more efficient than the ordinary version if the exponent needs to be larger than necessary from a security point of view (e.g. when a collision-free number generator is used).

In the remainder of this conclusion, some remarks are given concerning the use of cryptographic co-processors, further properties and future prospects.

4.6.1 Using Cryptographic Co-Processors

Cryptographic co-processors are widely used in practice to increase the speed of computations. Often, such co-processors have the disadvantage that bit-lengths are static. In the case of discrete-logarithm-based cryptosystems with the settings discussed in this section, this can happen for the lengths of q and p'. As motivated in the introduction, however, the subsequent application may require the use of longer secret keys due to coding reasons, rather than security reasons. Obviously, in the ordinary setting, where one would have to choose q sufficiently large, the cryptographic hardware cannot be used anymore if the supported bit-lengths are exceeded. However, if the secret key is larger than q, but smaller than $q2^{l_q} + q$, then by using $\mathbb{G}_p$ instead of $\mathbb{G}_q$, the cryptographic hardware can still be used. A similar argument holds for $\mathbb{G}_{\hat{n}}$.

4.6.2 Further Properties

In $\mathbb{G}_p$, inverting an element is normally done by component-wise multiplicative inversion in $\mathbb{G}_q$. This can be expressed through fusion-exponentiation as follows:

$$(a,b)^{-1} = (a^{-1}, b^{-1}) = (a^{-1}b^0, a^0b^{-1}) = (a,b)^{(-1,0)}$$

Multiplication in $\mathbb{F}_p$ produces interesting properties:

$$
\begin{array}{lll}
(a,b)(0,1) = (-b,a) & (a,b)(i,0) = (ai,bi) & (a,b)(i,i) = (i(a-b), i(a+b)) \\
(a,0)(1,1) = (a,a) & (0,a)(1,1) = (-a,a) & (a,0)(b,0) = (ab,0) \\
(a,0)(0,b) = (0,ab) & (0,a)(b,0) = (0,ab) & (0,a)(0,b) = (-ab,0)
\end{array}
$$

4.6.3 Future Prospects

Some interesting research that goes beyond the scope of this introductory work, is using the concept of fusion to define further variants of the Discrete Logarithm Problem Family. For instance, one could consider the case where components of exponents follow other, non-uniform probability distributions. This gives some further variants to analyze, especially because there are several combinations. For instance, consider a fusion-exponent (a,b), where a is uniformly distributed over $\mathbb{Z}_q$, but b is normally distributed over $\mathbb{Z}_q$. How difficult is it to solve the Fusion Discrete Logarithm Problem?

Intuitively, basing the RSA-cryptosystem on $\mathbb{G}_{\hat{n}}$ is secure with the assumption that factoring n is hard. Future work in this context may include a provable reduction and some analysis in combination with ElGamal run over $\mathbb{G}_{\hat{n}}$.

Distributed Generation of Unique Keys

Chapter 5

5.1 Introduction

The techniques presented in Chapter 3 can be used to locally generate system-wide unique secret keys. This is important in the context of public-key cryptography, since identical public keys can enable mutual impersonation for the corresponding owners (cf. Chapter 2.5). Many IT-systems require the existence of *one* trusted party to generate and certify keys. This, however, is an unrealistic requirement, because in practice no single trustworthy party exists. A typical example is a political election. There, the tally is not performed by just one single party. Instead, some kind of four-eyes principle is applied to ensure that the tally is done correctly.

A lot of research has focused on developing techniques to ensure that functions which are based on $\mathbb{Z}_q$, where q is a prime, can be shared securely among a set of n instances, called players. These techniques include secret sharing [Sha79, Fel87, Ped91b] and (secure) multi-party computation [GMW87, GRR98]. Most of the proposed protocols provide security up to a particular threshold t of passive or active adversaries. In the passive case this means that up to t players, who follow the specification, are not able to gain any information about the secret, whereas $t+1$ up to n players are able to perform a reconstruction. The same is true in the active case, however the players may behave dishonestly in *any* way, i.e. they can even deviate from the protocol specification.

A practical use of multi-party computation is the sharing of a public-key cryptosystem, which is then referred to as threshold cryptosystem [DF90]. Here, at least $t + 1$ players can jointly *generate* and *use* a secret key without any reconstructions. The shared generation of a secret key requires special protocols. For discrete-logarithm-based cryptosystems the protocols proposed in [Ped91a, GJKR99] are widely used. Such protocols can be used to distribute a certification authority: the players jointly generate the secret key for the user, but do not reconstruct it. Instead, they send the corresponding shares to the user, who preforms the reconstruction independently. The corresponding public key is reconstructed and jointly certified by the players. Asides from sharing a trust center, applications also exist where the secret key is never reconstructed, i.e. it is *used* in shared from. In the context of electronic elections, this is necessary to ensure that no single party is able to decrypt single votes. Another example is the shared generation of signatures, where several people or organizations have to agree on a content of a document and then jointly generate *one* signature. The techniques can also be used to hamper robbery in banks (several people have to agree to open a bank vault) or to prevent from unauthorized launching of nuclear missiles.

5.1.1   Collision-Free Distributed Key Generation

The current chapter addresses the problem of allowing a set of players to jointly generate a *system-wide unique* secret key without reconstructing intermediate results. The focus lies on distributed (or shared) key generation for discrete-logarithm-based threshold cryptosystems, especially ElGamal-like schemes [ElG85a, DF90]. In such schemes, a secret key z is chosen at random from $\mathbb{Z}_q$, where q is an odd prime. The corresponding public key h is defined as $h = g^z$, where g is a generator of $\mathbb{G}_q$.

To guarantee that a secret key z is system-wide unique, a collision-free number generator has to be designed with respect to the design principles stated in Chapter 3 and with the additional restriction $z < q$. Otherwise z is reduced modulo q and not necessarily unique anymore. The goal is to share the following computation:

$$f(u, r)2^{l_r} + r$$

In the threshold setting, r needs to be generated by a set of players such that it is restricted to the l_r least significant bits, while at the same time being uniformly distributed over $[0, 2^{l_r} - 1]$. Furthermore, f needs to be a function that can be shared in $\mathbb{Z}_q$ using secure multi-party computation and whose output can be shifted to the left of r whilst preserving $z < q$. Although this sounds easy, it is quite challenging to share such a construction while preserving the required uniform distribution of r. Alternatively, provably secure solutions exist for the shared generation of secrets that are uniformly distributed in $\mathbb{Z}_q$. Assume that the random part r, used for the collision-free number generator, has been generated using one of these solutions. Furthermore, let f be defined over $\mathbb{Z}_q$ exclusively, i.e. $f : \mathbb{Z}_q \times \mathbb{Z}_q \rightarrow \mathbb{Z}_q$. From Theorem 3.2.1 it follows that the *pair* $(f(u, r), r) \in \mathbb{Z}_q^2$ is system-wide unique if u is system-wide unique and can be computed by standard techniques of secure multi-party computation. The output of f and r remain shared. The use of standard techniques has the advantage that the above construction requires no (or minor) additional security proofs. Notice that the secret key z is a *pair* of integers. For such a representation, the concept of fusion, as introduced in Chapter 4, has to be used. This in turn requires the development of special protocols for secure multi-party computation in $\mathbb{F}_p$.

In this chapter, three different approaches are given:

1. *Shared Collision-Free Number Generation:* Each player holds a share of u, which is restricted to certain bit-positions. In each run of the protocol, the players jointly generate a random secret r which is also restricted to certain bit-positions. A suitable function f is evaluated over u and r using secure multi-party computation. The output of f is concatenated to r, i.e. $z = f(u,r)2^{l_r} + r$. The protocol is designed such that $z < q$ is preserved. In the end, each player holds a share of z.

2. *Shared Collision-Free Pair Generation:* Each player holds a share of $u \in \mathbb{Z}_q$. During a run of the distributed key generation protocol, the players jointly generate a random secret $r \in \mathbb{Z}_q$. Then, a function f, that is based on computations done uniformly in $\mathbb{Z}_q$ is evaluated using secure multi-party computation. In the end each player holds a share of each component of $\mathbf{z} = (f(u,r), r)$, where $\mathbf{z} \in \mathbb{F}_p$.

3. *Shared Multiplication of Unique Primes:* Each player P_i generates a prime p_i using a (local) collision-free number generator and shares it over the other players. Then, a shared multiplication protocol is performed over the shared primes. If the primes are restricted to certain bit-positions then their product $z = p_1 \cdot \ldots \cdot p_n$ is unique due to the Fundamental Theorem of Arithmetic. It is therefor necessary that $z < q$. In the end, each player holds a share of z.

Solutions 1 and 3 require that several involved secrets are restricted to certain bit-positions. The challenge for these solutions is to develop a protocol which provides correctness, sufficient secrecy (notice that keys are not uniformly distributed) and system-wide uniqueness of the generated keys. The second approach has the advantage that standard techniques for secure multi-party computation can be used, which simplifies the analysis with respect to several requirements. Each of the protocols is designed so that each player gets the public key as a by-product.

5.1.2   Contribution and Organization

The contribution of this chapter is the design of the three approaches mentioned above. The chapter is organized as follows. In Section 5.2, a large range of necessary preliminaries is given. Section 5.3 presents secret sharing and multi-party computation for $\mathbb{F}_p$ with respect to the fusion-concept. In Section 5.4.1, the standard requirements for distributed key generation are adapted to include system-wide uniqueness. The three approaches above stated are then sketched on an abstract level. Corresponding practical instantiations are specified and analyzed in Sections 5.5, 5.6, 5.7 and 5.8. In the end open problems and a comparison between the presented protocols are stated.

5.2   Preliminaries

Throughout this chapter, it is assumed that $\mathbb{G}_q$ is a group of prime order q, where the Discrete Logarithm Problem is believed to be hard (cf. Section 4.2.3). Furthermore, g, g_1 and g_2 are generators of $\mathbb{G}_q$, where the mutual discrete logarithms are unknown.

The following theorem is known as the Fundamental Theorem of Arithmetic [Bun02]:

Theorem 5.2.1 (Gauss)   *Every natural number z except 1 can be represented by the product $p_1^{e_1} \cdot \ldots \cdot p_l^{e_l}$, where $p_1, \ldots, p_l \in \mathbb{P}$, $e_1, \ldots, e_l \in \mathbb{N}$, $p_i < p_{i+1}$ for $i = 1, \ldots, l-1$, and $l \in \mathbb{N}$. Hereby, the sequence $p_1^{e_1}, \ldots, p_l^{e_l}$ is unique.*

The number of primes less than a given number $z \in \mathbb{N}$ can be approximated [Bun02]:

Theorem 5.2.2 (Riemann)   *Let z be a natural number. Then the number of primes smaller than z is approximately $\frac{z}{log(z)}$.*

5.2.1   Uncommon Discrete Logarithm Problems

In Section 4.2.3, the Discrete Logarithm Problem was defined for a group $\mathbb{G}_q$ of prime order q, where the exponent is implicitly assumed to be uniformly distributed over $\mathbb{Z}_q$. The latter is guaranteed if a random number generator is involved. In practice, however, exponents are often generated using a pseudo-random number generator. Moreover, it might have a particular structure due to coding reasons. In [Tes01], a survey is given of several exotic variants of the Discrete Logarithm Problem and the approximate running times of the corresponding best known attack algorithms.

5.2.1.1   Known Interval

The larger the order of a group is, the longer exponentiations last in average. For efficiency reasons, it is useful to use small exponents, but this can decrease the security of a system if the system parameters are not chosen carefully. Care is needed especially if the exponent is a secret key. This leads to the following definition:

Definition 5.2.1   Let $a, b \in \mathbb{Z}_q$ and $y = g^x$, where $x \in [a, b]$. The *Interval Discrete Logarithm Problem* (IDLP) is the following: given y, g, a, b and q, find x.

If $a = 0$ and $b = q - 1$, this is an instance of the ordinary Discrete Logarithm Problem. The best known attack against the IDLP is Pollard's kangaroo method [Pol78, Pol00], with an expected running time of $O(\sqrt{b-a})$.

5.2.1.2   Known Expected Value

Apart from knowing that a discrete logarithm lies in a certain interval, perhaps its expected value is known:

Definition 5.2.2 Let $y = g^x$, where $x \in \mathbb{Z}_q$ and $E(x)$ is the expected value of x. The *Expected Value Discrete Logarithm Problem* (EDLP) is the following: given y, g, $E(x)$ and q, find x.

If $E(x) = q/2$, this is an instance of the ordinary Discrete Logarithm Problem. The best known attack to solve the EDLP requires $O(\sqrt{E(x)})$ steps.

5.2.1.3   Known Probability Distribution

Definition 5.2.3 Let $y = g^x$, where x is chosen from $\mathbb{Z}_q$ with respect to the probability distribution (p_i), where $p_i = \Pr[x = i]$. The *Probability Distribution Discrete Logarithm Problem* (PDLP) is the following: given y, g, (p_i) and q, find x.

If $p_i = 1/q$ for all $0 \leq i < q$ and $p_i = 0$ otherwise, this is an instance of the ordinary Discrete Logarithm Problem. It seems that the known probability distribution cannot be used to optimize Pollard's rho algorithm [BT00]. If a secret follows the normal or geometric distribution, then the PDLP becomes the EDLP [BT00] if the baby-giant step algorithm is adapted. For further discussions of other distributions, such as the Pareto or Weibull distribution, see [BT00].

5.2.2   Homomorphic Commitments

Commitment schemes [BCC88, CD97, Dam99] are very important primitives to secure systems against active malicious participants. They are mainly used to identify misbehavior during cryptographic computations so that the correctness of the results can be guaranteed or at least so that wrong outputs are identified. A commitment scheme consists of the *commit-stage* and the *open-stage*. During the first stage, an instance generally commits to a secret through an injective one-way function that is believed to be computationally hard to invert. The output of the commit-function is called the *commitment* or *blob*, and generally sent to a different instance or just made public. Furthermore, there needs to exist some kind of *open information*, which allows the secret to be extracted from the commitment during the open-stage, or is at least sufficient to verify if a secret corresponds to a given commitment. For simplicity, an instance that commits to a secret is called a *prover* and an instance that receives, verifies or opens the commitment is referred to as a *verifier*. The security of a commitment scheme is given by two properties:

1. *Hiding:* The verifier gains no information about the secret before the open-stage.

2. *Binding:* The prover is not able to generate open-information which leads to the disclosure of a secret that differs from the one involved in the commit-stage.

Two assumptions follow accordingly for the prover and the verifier:

1. If the prover (verifier) is an unbounded algorithm, then binding (hiding) needs to hold *perfectly*, i.e. with information-theoretic security.

2. If the prover (verifier) is a poly-bounded algorithm, then binding (hiding) needs to hold *computationally*, i.e. with complexity-theoretic security.

Notice, however, that both properties cannot hold perfectly simultaneously, since it is impossible for a commitment to contain and not to contain information about the secret at the same time [Dam99].

Whether the open-stage is executed depends on the application for which the commitment scheme is used. Sometimes, it suffices to prove that a commitment is formed correctly and that it is only opened in the case that the prover behaves dishonestly in subsequent protocol-steps. Apart from the security properties, a commitment scheme must be provably correct, i.e. if the prover and the verifier behave honestly, then the verifier accepts both stages and receives the correct secret during the open-stage.

In the following, two discrete-logarithm-based commitment schemes are described. The first provides perfect binding and computational hiding, whereas the second provides computational binding and perfect hiding. Both schemes are based on group homomorphisms, which are intrinsic to achieve security against active adversaries.

5.2.2.1   Perfect Binding Commitments

Definition 5.2.4 Let $s \in \mathbb{Z}_q$ be a secret. The *Discrete Logarithm Commitment* (D-commitment) is given by $S = g^s$.

The open-information is the secret s. Let $\hat{s}$ be the conjectured open-information which the verifier has on hand. The open-stage succeeds if $S = g^{\hat{s}}$ holds. The correctness of the commit and open stages is self-evident. It remains to show the following:

Lemma 5.2.1 *D-commitments are (1) perfect binding and (2) computational hiding.*

Proof. (1) Since exponentiation in $\mathbb{G}_q$ is injective, there exist no two distinct elements $s, \hat{s} \in \mathbb{Z}_q$, such that $g^s = g^{\hat{s}}$ holds. (2) Computing s from S and g is equivalent to solving the Discrete Logarithm Problem in $\mathbb{G}_q$.   □

The commit-function is the exponentiation in $\mathbb{G}_q$, which is well known to be a group homomorphism from $(\mathbb{Z}_q, +)$ to $(\mathbb{G}_q, \cdot)$ for any (fixed) basis g.

5.2.2.2   Perfect Hiding Commitments

In [Ped91b], Pedersen proposed a discrete-logarithm-based commitment scheme which provides computational binding, but perfect hiding. Such a scheme is quite useful when the number of possible secrets is small.

Definition 5.2.5 Let $s \in \mathbb{Z}_q$ be a secret. The *Pedersen Commitment* (P-commitment) is given by $\widetilde{S} = g_1^s g_2^{s'}$, where $s' \in_R \mathbb{Z}_q$.

Hereby, the open-information is (s, s') and the open-procedure works straightforwardly, analogously to that of the D-commitment scheme. It is important that the prover does not know $dlog_{g_1}(g_2)$, since otherwise the binding-property can be broken:

Lemma 5.2.2 *P-commitments are (1) computational binding and (2) perfect hiding.*

Proof. (1) Let $\widetilde{S} = g_1^s g_2^{s'} = g_1^{s+ws'}$, where $w = dlog_{g_1}(g_2)$. To open $\widetilde{S}$ resulting in a value $\hat{s} \neq s$, the dishonest prover has to find an appropriate $\hat{s}'$, such that $\hat{s} + w\hat{s}' = s + ws'$. However, solving the Discrete Logarithm Problem is computationally equivalent to breaking the binding-property: if the prover can solve the Discrete Logarithm Problem, then he can obtain w and compute $\hat{s}' = w^{-1}(s - \hat{s}) + s'$. For the converse the existence of an oracle $\mathcal{O}$ is assumed that, on the input of $\widetilde{S}$, $\hat{s}$, g_1 and g_2, returns $\hat{s}'$, such that $\widetilde{S} = g_1^{\hat{s}} g_2^{\hat{s}'}$ holds. Let $y \in \mathbb{G}_q$ for which $dlog_g(y)$ needs to be obtained. Then $\mathcal{O}(g^0 y, 0, g, g)$ can be queried to get $dlog_g(y)$. (2) From Lemma 5.2.1, a secret s is bound perfectly by a D-commitment g_1^s. Since $s' \in_R \mathbb{Z}_q$, $g_2^{s'}$ is a random blinding for g_1^s and so every element of $\mathbb{G}_q$ is a valid P-commitment for every s. □

For any (fixed) pair (g_1, g_2), the commit-function is a group homomorphism from $(\mathbb{Z}_q^2, +)$ to $(\mathbb{G}_q, \cdot)$, where $+$ denotes the component-wise addition. $(\mathbb{Z}_q^2, +)$ is an abelian group with the neutral element $(0, 0)$. Let $(a, a'), (b, b') \in \mathbb{Z}_q^2$, $\widetilde{A} = g_1^a g_2^{a'}$ and $\widetilde{B} = g_1^b g_2^{b'}$. For $(a, a') + (b, b') = (a + b, a' + b')$ one gets the commitment $g_1^{a+b} g_2^{a'+b'}$ which is equivalent to $\widetilde{A}\widetilde{B} = g_1^a g_2^{a'} g_1^b g_2^{b'} = g_1^{a+b} g_2^{a'+b'}$. The inverse in $\mathbb{Z}_q^2$ maps onto the inverse in $\mathbb{G}_q$, since $-(a, b) = (-a, -b)$ and $g_1^{-a} g_2^{-b} = (g_1^a g_2^b)^{-1}$. Moreover, the neutral element in $(\mathbb{Z}_q^2, +)$ maps onto the neutral element in $\mathbb{G}_q$, i.e. $g_1^0 g_2^0 = 1$.

5.2.3   Σ-Protocols

Sometimes commitments do not sufficiently secure a system against active adversaries. A far more powerful technique is an interactive proof-system, introduced in [GMR89] (cf. [Gol02] for a good overview). This work focuses on special kinds of such proof-systems called Σ-*protocols* (or Σ-proofs) [Cra96]. Such protocols involve the exchanging of three messages between a prover and a verifier. The prover wants to convince the verifier that it knows a secret witness w, satisfying a public predicate Q, without revealing any information about it. At the beginning of a Σ-protocol (cf. Figure 5.1), the prover sends the first-message t to the verifier, which is generated using randomness and possibly information about w. After having received t, the verifier chooses a challenge c from a particular space and returns it to the prover. Finally, the prover sends a response s to the verifier, whose generation involves c, information for the generation of t and possibly information about w. The verifier accepts a protocol run if (t, c, s) is formed correctly with respect to Q. If the prover is able to guess the correct challenge, then he can cheat the verifier. So, the protocol may need to be repeated several times until the probability of the prover successfully cheating is negligible.

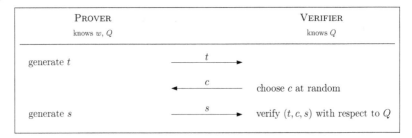

Fig. 5.1: Structure of a Σ-Protocol.

A Σ-protocol requires that the following properties hold:

P1 (Completeness): If the prover and the verifier follow the specified protocol steps, then the verifier is convinced that the prover knows a witness w satisfying Q.

P2 (Special Soundness): A poly-bounded algorithm (the *knowledge-extractor*) must exist which on the input of two protocol triples (t, c, s) and (t, c^*, s^*), where $c \neq c^*$, is able to compute the witness w efficiently.

P3 (Special Honest-Verifier Zero-Knowledge): A poly-bounded algorithm (the *simulator*) must exist which, given any challenge c, generates t' and s' such that (t', c, s') is a valid protocol triple. Moreover, (t', c, s') must have the same probability distribution as a real run of the protocol. Since the verifier is assumed to be honest, he chooses the challenge of each run at random (i.e. independent of previous rounds). A cheating verifier can generate challenges dependent on prior rounds of the protocol. If a simulator can be given even for this setting, then the protocol is called a perfect zero-knowledge protocol. However, the size of the challenge space needs to be poly-bounded. This restriction is not necessary for the setting where the verifier is assumed to be honest. In this case, the challenge space can be even exponential, which improves the efficiency in terms of the number of exchanged messages.

In practice, it is not advisable to assume that a verifier is honest. Fortunately, using a simple trick, Σ-protocols can be used in practice, even in the presence of a malicious verifier: using random oracles, the verifier can be "forced" to behave honestly.

5.2.3.1 Non-Interactive Σ-Protocols

Interactive proofs can also be made non-interactive [BFM88, LS91]. Here, the prover generates only one message that is sufficient to convince the verifier. A practical approach, known as the Fiat-Shamir paradigm, has been suggested in [FS87]. The idea is that any interactive proof of knowledge (in particular a Σ-protocol) can be turned into a digital signature scheme. However, it is important to ensure that the challenge is chosen *after* the computation of the first message, as is the case in a real protocol run. If this can be guaranteed although the prover does it by himself, then a verifier must accept the proof and cannot cheat the prover, because he cannot influence the generation of c. In [FS87], it has been suggested to set c as the output of a cryptographic hash function $\mathcal{H}$ applied to the first-message. Since no interaction with the verifier is required, a Σ-protocol is then zero-knowledge even if the verifier is malicious. It is hardly possible to prove the security of such a solution due to the involvement of hash functions. Thus, security proofs in this context assume the existence of an ideal hash function. The security is then said to hold in the *random oracle model* [BR93, PS96]. Further interesting discussions related to the Fiat-Shamir paradigm can be found in [AABN02, MR02].

Example 5.2.1 Let $y = g^x$, where $x \in \mathbb{Z}_q$. The prover wants to prove knowledge of x without interaction [Sch89, Sch91]: first he computes $t = g^\alpha$, where $\alpha \in_R \mathbb{Z}_q$. Then he sets $c := \mathcal{H}(t)$, where $\mathcal{H} : \{0,1\}^* \to \mathbb{Z}_q$ is a cryptographic hash function, and finally computes $s = \alpha - cx$. The non-interactive proof is then the pair (c, s). Given (c, s), any verifier that has access to $\mathcal{H}$, y and g can check if $c = \mathcal{H}(y^c g^s)$ holds. If the prover includes a message m when hashing t, then (c, s) is a digital signature with respect to m and the signature key x.

Henceforth, non-interactive proofs are used exclusively in this chapter as the description of the protocols is more concise and the protocols are then even secure against (active) dishonest verifiers. We call the pair (c, s) the non-interactive proof and denote it by $\tau = (\tau_1, \tau_2) := (c, s)$. The verification procedure is given by the black-box function V. In Example 5.2.1, $V : T \times \mathbb{G}_q \to \{0, 1\}$ would be defined as follows:

$$V(\tau, y) := \begin{cases} 0 & \text{if } \tau_1 \neq \mathcal{H}(y^{\tau_1} g^{\tau_2}) \\ 1 & \text{otherwise} \end{cases}$$

where T denotes the set of non-interactive proofs. Notice that in [GK03], the Fiat-Shamir paradigm has been shown to possibly lead to an insecure non-interactive proof, although the corresponding interactive one has been proven to be secure. [GK03] is recommended for further notes.

In the following, two Σ-Protocols are given. The first is necessary to prove that a triple $(\widetilde{X}, \widetilde{Y}, \widetilde{Y})$ is of the form $(g_1^x g_2^{x'}, g_1^y g_2^{y'}, g_1^{xy} g_2^{x'y'})$. The second is a Σ-OR-proof which can prove that for a given P-commitment $\widetilde{X}_i$, the corresponding secret is either 0 or 2^i, for $2^i < q$. Using several instances of such a proof, it can be proven that the secret x contained in a P-commitment $\widetilde{X}$ lies in a specific interval.

5.2.3.2   Proving that a Triple is a Pedersen-Diffie-Hellman-Triple

Let $\widetilde{X}$ and $\widetilde{Y}$ be two P-commitments of the secrets x and y. Assume that the specification requires the prover to open the commitment of the sum $z = x + y$. A verifier can verify the open-process, since he can obtain $\widetilde{Z}$ (the P-commitment of z) by computing $\widetilde{X}\widetilde{Y}$. This is possible, because the commit-function is a group homomorphism. Sometimes, it might be the case that the specification requires the prover to open the commitment of the product $z = xy$. If the Diffie-Hellman Problem is hard in $\mathbb{G}_q$, then it is infeasible for a poly-bounded verifier to efficiently compute the corresponding commitment $\widetilde{Z}$ from $\widetilde{X}$ and $\widetilde{Y}$. Hence, the prover has to send $\widetilde{Z}$ to the verifier and convince him that $\widetilde{Z}$ is indeed a commitment of xy. To simplify further discussions, the following definition is introduced:

Definition 5.2.6 Let $\widetilde{X} = g_1^x g_2^{x'}$ and $\widetilde{Y} = g_1^y g_2^{y'}$, where $x, x', y, y' \in \mathbb{Z}_q$. If $\widetilde{Z} = g_1^{xy} g_2^{x'y'}$ holds $(\widetilde{X}, \widetilde{Y}, \widetilde{Z})$ is called *Pedersen-Diffie-Hellman-Triple*.

In [CD98], a general method is given for proving that a given commitment contains a secret which is the product of two other committed secrets. A special variant for P-commitments can be found in [GRR98]. The Σ-protocol in Figure 5.2 is a slightly modified variant of the approach in [GRR98]. In this case, $\widetilde{Z} = g_1^{xy} g_2^{x'y'}$, whereas in [GRR98], $\widetilde{Z} = g_1^{xy} g_2^{z'}$, for some $z' \in_R \mathbb{Z}_q$.

The idea behind the protocol in Figure 5.2 is that $\widetilde{Z}$ can be transformed as follows:

$$\widetilde{Z} = g_1^{xy} g_2^{x'y'} = g_1^{xy} g_2^{x'y} g_2^{x'y'-x'y} = (g_1^x g_2^{x'})^y g_2^{x'(y'-y)} = \widetilde{X}^y g_2^{x'(y'-y)}$$

The prover has to prove that $\widetilde{Z}$ is of the above form. Since x', y' and y are involved, he must additionally prove that he knows the representation of $\widetilde{X}$ and $\widetilde{Y}$. The necessary techniques are well known and include the proofs for the knowledge [Sch89] and equality [CP93] of discrete logarithms.

Theorem 5.2.3 *The Σ-protocol in Figure 5.2 is a proof that $(\widetilde{X}, \widetilde{Y}, \widetilde{Z})$ is a Pedersen-Diffie-Hellman-Triple with respect to Definition 5.2.6. The proof satisfies (1) completeness, (2) special soundness, and (3) is special honest-verifier zero-knowledge.*

Proof. It has to be shown that (1), (2) and (3) hold.

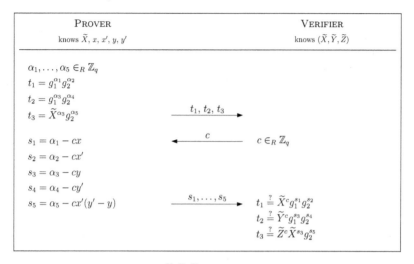

Fig. 5.2: Σ-proof, that $(\widetilde{X}, \widetilde{Y}, \widetilde{Z})$ is a Pedersen-Diffie-Hellman-Triple.

(1) If the prover and the verifier follow the specification, then the verification process succeeds which is shown in the following:

$$\widetilde{X}^c g_1^{s_1} g_2^{s_2} = g_1^{xc} g_2^{x'c} g_1^{\alpha_1-cx} g_2^{\alpha_2-cx'} = g_1^{\alpha_1} g_2^{\alpha_2} = t_1$$
$$\widetilde{Y}^c g_1^{s_3} g_2^{s_4} = g_1^{yc} g_2^{y'c} g_1^{\alpha_3-cy} g_2^{\alpha_4-cy'} = g_1^{\alpha_3} g_2^{\alpha_4} = t_2$$
$$\widetilde{Z}^c \widetilde{X}^{s_3} g_2^{s_5} = g_1^{xyc} g_2^{x'y'c} g_1^{x(\alpha_3-cy)} g_2^{x'(\alpha_3-cy)} g_2^{\alpha_5-cx'(y'-y)}$$
$$= g_1^{xyc} g_2^{x'y'c} g_1^{x\alpha_3-xcy} g_2^{x'\alpha_3-x'cy} g_2^{\alpha_5-cx'y'+cx'y}$$
$$= g_1^{x\alpha_3} g_2^{x'\alpha_3} g_2^{\alpha_5} = \widetilde{X}^{\alpha_3} g_2^{\alpha_5} = t_3$$

(2) Let (t, c, s) and (t, c^*, s^*) be two conversation triples, where $t = (t_1, t_2, t_3)$, $c \neq c^*$, $s = (s_1, \ldots, s_5)$ and $s^* = (s_1^*, \ldots, s_5^*)$. Let $a_1 = x$, $a_2 = x'$, $a_3 = y$, $a_4 = y'$ and $a_5 = x'(y'-y)$. Since $c^* - c$ is co-prime to q, each a_i can be extracted as follows:

$$(s_i - s_i^*)(c^* - c)^{-1} = (\alpha_i - ca_i - \alpha_i + c^* a_i)(c^* - c)^{-1} = a_i(c^* - c)(c^* - c)^{-1} = a_i$$

(3) Let (t, c, s) be the transcript of a real protocol run. The simulated triple (t', c, s') for a correct conversation is computed as follows:

1. Choose $s_1', \ldots, s_5' \in_R \mathbb{Z}_q$ and set $s' = (s_1', \ldots, s_5')$.
2. Compute $t_1' = \widetilde{X}^c g_1^{s_1'} g_2^{s_2'}$, $t_2' = \widetilde{Y}^c g_1^{s_3'} g_2^{s_4'}$, $t_3' = \widetilde{Z}^c \widetilde{X}^{s_3'} g_2^{s_5'}$ and set $t' = (t_1', t_2', t_3')$.

The first-messages t and t' are defined over (c, s) and (c, s'). Let C, S and S' be random variables for c, s and s'. It remains to be shown, that the conditional distributions $\Pr[S|C = c]$ and $\Pr[S'|C = c]$ are identical. Due to step 1 of the simulator, each s'_i is uniformly distributed in $\mathbb{Z}_q$. In a real protocol run, each $\alpha_i \in \mathbb{Z}_q$ is also chosen with uniform distribution. All computations are done in $\mathbb{Z}_q$ and so each s_i is uniformly distributed in $\mathbb{Z}_q$. Hence, the distributions are indistinguishable.   □

The Σ-protocol in Figure 5.2 can be made non-interactive: the prover computes $t_1, \ldots, t_3$ and $s_1, \ldots, s_5$ as stated in Figure 5.2. He sets $c := \mathcal{H}(t_1 || t_2 || t_3)$ and $\tau := (c, s_1, \ldots, s_5)$. Given τ and the triple $(\widetilde{X}, \widetilde{Y}, \widetilde{Z})$, one can verify the correctness of τ with the function $V : T \times \mathbb{G}_q^3 \to \{0, 1\}$, which is defined as follows:

$$V(\tau, (\widetilde{X}, \widetilde{Y}, \widetilde{Z})) := \begin{cases} 0 & \text{if } \tau_1 \neq \mathcal{H}(\widetilde{X}^{\tau_1} g_1^{\tau_2} g_2^{\tau_3} || \widetilde{Y}^{\tau_1} g_1^{\tau_4} g_2^{\tau_5} || \widetilde{Z}^{\tau_1} \widetilde{X}^{\tau_4} g_2^{\tau_6}) \\ 1 & \text{otherwise} \end{cases}$$

5.2.3.3   Proving that a Discrete Logarithm lies within a specific Interval

For a P-commitment $\widetilde{X}$ it is straightforward to run a proof of knowledge for a committed secret $x \in \mathbb{Z}_q$. In some situations, however, the specification may require that x is an element of a particular set $M \subseteq [0, q-1]$. In addition to proving knowledge of x, the prover then has to show that x lies within M. In [Mao98], a solution has been given for the case that $M = [0, 2^{l_k} - 1]$, where $2^{l_k} - 1 < q$. There, the prover must show that the bits at the positions l_k up to $l_q - 1$ are equal to 0 and that he knows all other bits. In the following, this approach is generalized to work for the set

$$M = \left\{ \sum_{i \in B_1} 2^i + \sum_{j \in B} x_j 2^j \;\middle|\; x_j \in \{0, 1\} \;\wedge\; B_0 \cup B_1 \cup B = [0, l_q - 1] \right\}.$$

Here, B_i is the set of bit-positions where the bits are always equal to i. The set B contains the bit-positions which are either 0 or 1. B, B_0 and B_1 are pair-wise disjoint, public and authentic, i.e. can be contained in the system parameters. The protocol given in this section ensures that the prover knows all the bits at the positions in B of a secret x that corresponds to the P-commitment $\widetilde{X}$. Moreover, the proof ensures that all other bits are set according to B_0 and B_1. If the proof succeeds, $\widetilde{X}$ is a P-commitment of $x \in M$. The description of the protocol can be simplified by making the following definitions with respect to B, B_1 and $\widetilde{X} = g_1^x g_2^{x'}$:

$$b_i = g_1^{2^i}, \quad \widehat{B}_1 = \prod_{i \in B_1} b_i, \quad \widetilde{X}_b = \prod_{i \in B} \widetilde{X}_i, \quad \widetilde{X}_i = b_i^{x_i} g_2^{x'_i}, \quad x' = \sum_{i \in B} x'_i$$

where $x = \sum_{i \in B_1} 2^i + \sum_{j \in B} x_j 2^j$, $x_j \in \{0, 1\}$ and $x'_i \in_R \mathbb{Z}_q$. Thus, the following holds:

$$\widetilde{X} = \widehat{B}_1 \widetilde{X}_b \qquad\qquad (5.2.1)$$

The correctness of Equation (5.2.1) can be seen from the following:

$$\widehat{B}_1 \widetilde{X}_b = \left(\prod_{i \in B_1} b_i \right) \left(\prod_{j \in B} \widetilde{X}_j \right) = \left(\prod_{i \in B_1} g_1^{2^i} \right) \left(\prod_{j \in B} b_j^{x_j} g_2^{x_j'} \right)$$

$$= g_1^{\sum_{i \in B_1} 2^i} \left(\prod_{j \in B} g_1^{x_j 2^j} \right) g_2^{\sum_{j \in B} x_j'}$$

$$= g_1^{\sum_{i \in B_1} 2^i + \sum_{j \in B} x_j 2^j} g_2^{x'} = g_1^x g_2^{x'} = \widetilde{X}$$

Prior to starting the interactive protocol shown in Figure 5.3, the prover needs to know x and its bit-representation, x' and B. The verifier needs to know $\widetilde{X}$, $\widehat{B}_1$ and B. At the beginning, the prover computes a P-commitment $\widetilde{X}_i = b_i^{x_i} g_2^{x_i'}$ of every bit x_i, whose position is contained in B. Hereby, he chooses each x_i' at random and with the condition $x' = \sum_{i \in B} x_i'$. Then he sends $\{\widetilde{X}_i\}_{i \in B}$ to the verifier, who computes $\widetilde{X}_b$ and checks if Equation (5.2.1) holds. If this is the case, the prover must prove that each $\widetilde{X}_i$ has been formed correctly, i.e. is a P-commitment of 0 or 2^i. The necessary Σ-proof is given at the bottom of Figure 5.3 and needs to be run for every index $i \in B$. Since the Σ-OR-Proofs are mutually independent, they can be run in parallel.

Theorem 5.2.4 *The Σ-protocol at the bottom of Figure 5.3 is a proof that $\widetilde{X}_i$ is a P-commitment of 0 or 2^i. The proof satisfies (1) completeness, (2) special soundness, and (3) is special honest-verifier zero-knowledge.*

Proof. It has to be shown that (1), (2) and (3) hold.

(1) If prover and verifier follow the specification, then the verification process succeeds which is shown in the following:

$$c_{i1} + c_{i2} = \begin{cases} (c_i - c_{i2}) + c_{i2} = c_i & \text{if } x_i = 0 \\ c_{i1} + (c_i - c_{i1}) = c_i & \text{if } x_i = 1 \end{cases}$$

$$\widetilde{X}_i^{c_{i1}} g_2^{s_{i1}} = \begin{cases} g_2^{x_i' c_{i1}} g_2^{\alpha_i - c_{i1} x_i'} = g_2^{\alpha_i} = t_{i1} & \text{if } x_i = 0 \\ t_{i1} & \text{if } x_i = 1 \end{cases}$$

$$(\widetilde{X}_i b_i^{-1})^{c_{i2}} g_2^{s_{i2}} = \begin{cases} t_{i2} & \text{if } x_i = 0 \\ (b_i^{c_{i2}} g_2^{x_i' c_{i2}} b_i^{-c_{i2}}) g_2^{\alpha_i - c_{i2} x_i'} = g_2^{\alpha_i} = t_{i2} & \text{if } x_i = 1 \end{cases}$$

(2) A knowledge extractor can be given straightforwardly for the non-simulated responses, i.e. if $x_i = 0$ then for s_{i1}, and if $x_i = 1$ then for s_{i2}.

(3) Let (t_i, c_i, s_i) be the transcript of a real protocol run. The simulated triple (t_i', c_i, s_i') for a correct conversation is computed as follows:

    1. Choose $s_{i1}', s_{i2}' \in_R \mathbb{Z}_q$ and set $s_i' := (s_{i1}', s_{i2}')$.

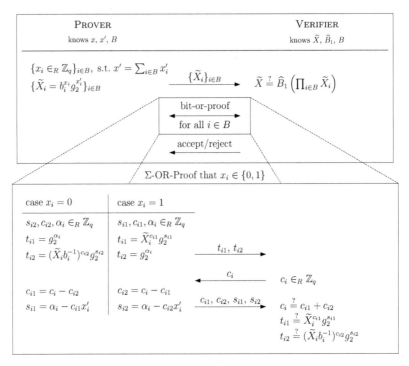

Fig. 5.3: Interactive Protocol to prove that $\widetilde{X}$ is a P-commitment of a Secret $x \in M$.

2. Compute $t'_{i1} = \widetilde{X}_i^{c_{i1}} g_2^{s'_{i1}}$, $t'_{i2} = (\widetilde{X}_i b_i^{-1})^{c_{i2}} g_2^{s'_{i2}}$ and set $t'_i = (t'_{i1}, t'_{i2})$.

The first-messages t_i and t'_i are defined over (c_{i1}, c_{i2}, s_i) and (c_{i1}, c_{i2}, s'_i). Let C, S and S' be random variables for (c_{i1}, c_{i2}), (s_{i1}, s_{i2}) and (s'_{i1}, s'_{i2}). It must be shown that $\Pr[S|C = c]$ and $\Pr[S'|C = c]$ are identical. The simulated part of the real protocol is obviously indistinguishable from the simulated transcript, since they are computed in the same way. It remains to consider the real part of the real protocol run. Since α_i is uniformly distributed in $\mathbb{Z}_q$ and all computations are done modulo q, it follows that s_{i1} (if $x_i = 0$) and s_{i2} (if $x_i = 1$) are also uniformly distributed in $\mathbb{Z}_q$. Hence, the distributions are indistinguishable.   □

Parallel executions of the Σ-OR-proof can be given through one non-interactive protocol: the prover computes $\{t_{i1}, t_{i2}\}_{i \in B}$ and $\{c_{i1}, c_{i2}, s_{i1}, s_{i2}\}_{i \in B}$ as stated in Figure 5.3, and sets $c := \mathcal{H}([t_{j1}||t_{j2}]_{j \in B})$ and $c_i := c$. Here, $[t_{j1}||t_{j2}]_{j \in B}$ denotes the concatenation of all blocks $t_{j1}||t_{j2}$. Finally, the prover sets $\tau := (c, \{(c_{i1}, c_{i2}, s_{i1}, s_{i2})\}_{i \in B})$. Given τ

and $\widetilde{X}$, τ can be verified using $V : T \times \mathbb{G}_q^{|B|} \to \{0,1\}$, defined as follows:

$$V(\tau, \{\widetilde{X}_j\}_{j\in B}) := \begin{cases} 0 & \text{if } \tau_1 \neq c_{i1} + c_{i2} \text{ or } \tau_1 \neq \mathcal{H}([\widetilde{X}_j^{c_{j1}} g_2^{s_{j1}} \| (\widetilde{X}_j b_j^{-1})^{c_{j2}} g_2^{s_{j2}}]_{j\in B}), \\ & \text{for any } (c_{i1}, c_{i2}, s_{i1}, s_{i2}) \in \tau_2, i \in B \\ 1 & \text{otherwise} \end{cases}$$

Notice, that because P-commitments are used, given $\widetilde{X}_i$ one cannot decide whether $x_i = 0$ or $x_i = 1$ with a chance better than $1/2$. Having B and B_1, the only information that an attacker gets, is that every bit x_i, where $i \in B_1$, is equal to 1 and every x_j, where $j \in B_0$ is equal to 0.

Example 5.2.2 Assume a P-commitment $\widetilde{S} = g_1^s g_2^{s'}$, where s is uniformly distributed over $[0, 2^{l_s} - 1] \subset [0, q-1]$, and l_s is small per assumption. Furthermore, the specification requires the computation of a D-commitment $S = g^s$, but l_s is so small that one can easily determine s from S simply using a brute force attack. To overcome this drawback, s can be padded by a random block ρ up to the length of q, such that $\rho 2^{l_s} + s < q$ is preserved. First, the holder of s chooses $\rho \in_R [0, 2^{l_\rho} - 1]$, such that $\rho 2^{l_s} + s < q$ holds, $\rho' \in_R \mathbb{Z}_q$ and then computes the P-commitment $\widetilde{\rho} = g_1^{\rho 2^{l_s}} g_2^{\rho'}$. Using the protocol stated in Figure 5.3 he then proves that $\rho \in \{x \mid x = x' 2^{l_s} \wedge x' \in \{0,1\}^{l_\rho}\}$. Thus, $B = \{i \mid l_s \leq i < l_s + l_\rho\}$ and $\widehat{B}_1 = 1$. If the proof is correct, *any* instance of the system can update $\widetilde{S}$ as follows:

$$\widetilde{S} = \widetilde{S}\widetilde{\rho} = (g_1^s g_2^{s'})(g_1^{\rho 2^{l_s}} g_2^{\rho'}) = g_1^{\rho 2^{l_s}+s} g_2^{s'+\rho'}$$

The holder of s then publishes the D-commitment $S = g^{\rho 2^{l_s}+s}$ and proves in zero-knowledge that $dlog_g(S) = dlog_{g_1}(\widetilde{S} g_2^{-s'-\rho'})$. Given $g^{\rho 2^{l_s}+s}$ and g, it is now harder to determine s than if g^s and g are given respectively.

By setting $B = [0, l_k - 1]$ for $2^{l_k} - 1 < q$ and $B_1 = \emptyset$, the solution in [Mao98] becomes a special case of our approach apart from a few slight differences: in [Mao98] the prover commits to x_i instead of $x_i 2^i$. Moreover, he chooses each x_i' at random, with the condition $x' = \sum_{i\in B} x_i' 2^i$. As a consequence, the verifier has to check if $\widetilde{X} = \prod_{i\in B} \widetilde{X}_i^{2^i}$ holds. Then, for each $\widetilde{X}_i$, the prover shows that it is a P-commitment of 0 or 1. Although the approach in [Mao98] achieves the same goal as the variant presented here, the protocol in Figure 5.3 is a little more efficient: $\widetilde{X} = \prod_{i\in B} \widetilde{X}_i$ is checked instead of $\widetilde{X} = \prod_{i\in B} \widetilde{X}_i^{2^i}$.

Remark 5.2.1 As none of the protocols presented later in this chapter makes use of verifiably setting specific bits to 1, it is henceforth assumed that $B_1 = \emptyset$.

5.2.4   Secret Sharing

Security infrastructures rarely exist without a trusted party. However, it is dangerous to place complete trust on one single instance. Once it has been corrupted, parts of or even the whole system becomes vulnerable. For instance, consider political elections. If ballots were tallied only by one authority, its corruption would lead to an undemocratic result. The same problem arises if elections are done electronically. Another famous area where techniques to share trust are intrinsic is the launching of nuclear missiles. Normally, several people have to agree at the same time to start the countdown. Sharing trust can also be helpful in preventing robberies: the key to open a vault can be shared among a set of employees.

The above examples show that there is a strong need to share trust among a set of parties so that only a qualified subset can simulate the tasks of one trusted party.

Sharing trust electronically means that some secret information, which is normally held by one instance, is shared among a set $\mathcal{P}$ of n instances (called players). Either all members of $\mathcal{P}$ or at least a qualified subset $\mathcal{Q}$ must be sufficient to reconstruct the secret information. Solutions which provide such properties are called *secret sharing schemes* [Sha79, Bla79]. For such schemes, two protocols are necessary [PHS03]:

1. *Sharing Protocol:* A player (the dealer) shares the secret over the players in $\mathcal{P}$.

2. *Reconstruction Protocol:* The players of $\mathcal{Q} \subseteq \mathcal{P}$ jointly reconstruct the secret.

In associated literature, the protocols are sometimes called the dealer-protocol and the combiner-protocol. Several solutions related to this topic can be found in [Sha79, Bla79, AB83, Sim91, Cac95, HJKY95] and [Sta96]. A widely used paradigm is *threshold secret sharing*, where a threshold t denotes the number of tolerable malicious players, i.e. up to t cooperating players are not able to (efficiently) obtain information about the secret. If more than t players agree to cooperate, correct reconstruction is easy. Two kinds of malicious players generally exist:

1. *Passive Adversaries:* Such adversaries at most try to learn something new from the protocol steps, but at least follow the specification.

2. *Active Adversaries:* Such adversaries are allowed to behave dishonestly in any way.

If up to t players are expected to be passively corrupt, then the secret sharing scheme does not need to contain any verification steps. If active adversaries are expected, then protocol steps need to be made verifiable so that misbehavior can be identified. Such extended schemes are called *verifiable secret sharing schemes* [CGMA85] and often use homomorphic commitments and interactive proofs as building blocks.

In the remainder of this section, basic schemes are described, which are secure against $t < n$ passive adversaries. Here, sum-sharing (with $t = n - 1$) and poly-sharing (with $t < n$) techniques are described, the latter also known as Shamir's secret sharing [Sha79]. Two sharing schemes are then examined which are secure against $t < n/2$ active adversaries, both of which are an extension of Shamir's scheme. Throughout this chapter, secret sharing is considered for secrets in $\mathbb{Z}_q$, where q is an odd prime.

5.2.4.1   Secret Sharing with Security against Passive Adversaries

Sum-Sharing: A simple way to share a secret $s \in \mathbb{Z}_q$ is to use *sum-sharing* [KGH83]. When sharing s, the dealer selects $n - 1$ random shares (sometimes called sum-shares) $s_1, \ldots, s_{n-1}$ in $\mathbb{Z}_q$ and computes $s_n = s - \sum_{i=1}^{n-1} s_i$. Hence, s is uniquely defined by all n shares. Having up to $n - 1$ shares obviously reveals *no* information about s, since computations are done modulo q. Although this approach is very simple, it has the disadvantage that *all* n shares are necessary to reconstruct the secret.

Shamir's Secret Sharing: In [Sha79], Shamir introduced one of the first threshold secret sharing schemes. Here, the dealer forms a t-degree polynomial as follows:

$$g(x) = s + \sum_{i=1}^{t} r_i x^i, \quad r_i \in_R \mathbb{Z}_q$$

where $s \in \mathbb{Z}_q$ is the secret. Then the i-th share (sometimes called poly-share) is obtained by evaluating $g(i)$. Any value in $\mathbb{Z}_q$ that is publicly and uniquely assigned to player P_i, can be used in the place of i, but this complicates a concise description. The i-th share is denoted by $[s_i]^q$ to avoid confusion. Since s and the coefficients are secret, $t + 1$ shares are necessary to reconstruct the secret using Lagrange interpolation:

$$s = \sum_{i=1}^{n} [s_i]^q \lambda_i, \quad \lambda_i = \prod_{\substack{j=1 \\ j \neq i}}^{n} j(j - i)^{-1}$$

Notice that it suffices to have $t + 1$ shares to successfully interpolate s. This leads to the following well known result (without proof):

Theorem 5.2.5 *Shamir's secret sharing scheme provides unconditional security if at most $t < n$ players are passively dishonest.*

Remark 5.2.2 Unconditional security requires that the messages sent over the network, are encrypted with an unconditionally secure encryption scheme [Sha49]. If a public-key infrastructure is used, then the system can only withstand bounded adversaries, because an unbounded adversary, that takes control over t players, can break one further channel and reconstruct the secret.

5.2.4.2   Secret Sharing with Security against Active Adversaries

The two schemes described in the previous section suffer from the problem that participants who do not follow the specification, cannot be identified. The worst anticipated behavior is that during the reconstruction stage a player agrees to participate but then sends an incorrect share, forcing a wrong output. In principle, there are the following situations where a player can send invalid information:

- *Sharing Stage:* The dealer distributes one or more invalid shares.

- *Reconstruction Stage:* One or more players input an invalid share.

By slightly modifying Shamir's scheme cheating players can be identified. In the following, two verifiable secret sharing schemes are described which require the use of homomorphic commitments and the existence of a broadcast channel. Moreover, it is assumed that adversaries are *static*, i.e. they decide on their behavior at the beginning.

Feldman's Verifiable Secret Sharing: In [Fel87], Feldman extended poly-sharing, providing security against $t < n/2$ active adversaries. Consider the share $[s_i]^q$ computed according to Shamir's scheme, i.e. $[s_i]^q = s + \sum_{j=1}^{t} r_j i^j$. Computing a D-commitment of $[s_i]^q$ player P_i obtains

$$S_i = g^{[s_i]^q} = g^{s + \sum_{j=1}^{t} r_j i^j} = g^s \prod_{j=1}^{t} \underbrace{(g^{r_j})}_{=:R_j}{}^{i^j} = g^s \prod_{j=1}^{t} R_j^{i^j}.$$

Henceforth, $\mathcal{T} := \{0, \ldots, t\}$ for simplicity. Let $r_0 = s$ and $R_0 = g^{r_0}$. If each player P_i knows $\{R_j\}_{j \in \mathcal{T}}$, then he can verify if $[s_i]^q$ is a node of the sharing-polynomial:

$$g^{[s_i]^q} \overset{?}{=} \prod_{j \in \mathcal{T}} R_j^{i^j} \tag{5.2.2}$$

Since the verification of every share requires the *same* set of commitments, the dealer only has to broadcast $\{R_j\}_{j \in \mathcal{T}}$ at the beginning of the distribution-stage, requiring the existence of a broadcast channel. If the verification fails for any player P_i, a complaint against the dealer is broadcasted. If there are more than t complaints, the dealer is disqualified. Disqualification means that each player forms the set $\mathcal{Q} \subseteq \mathcal{P}$ where only honest players are contained. If there are less than t complaints, the dealer has to broadcast $[s_i]^q$ for each complaining P_i, so that each player can verify the validity of Equation (5.2.2). If the verification fails, the dealer is disqualified, otherwise P_i.

For a correct reconstruction of s, the correctness of each submitted share needs to be verified. Therefore, the combiner performs the check stated in (5.2.2) for each involved $[s_i]^q$. He then performs the Lagrange interpolation over at least $t + 1$ valid shares. Notice that $t < n/2$ needs to hold, such that $t + 1$ honest players are still

Protocol PVSS

Input (Dealer): $s \in \mathbb{Z}_q$

Output (P_i): $([s_i]^q, [s_i']^q)$, $\{\widetilde{S}_j\}_{j \in I(\mathcal{Q})}$

1. The dealer chooses $s' \in_R \mathbb{Z}_q$ and forms the polynomials $g(x)$ and $g'(x)$ as follows:

$$g(x) = \sum_{i \in T} r_i x^i, \quad g'(x) = \sum_{i \in T} r_i' x^i$$

where $r_0 := s$, $r_0' := s'$ and $r_j, r_j' \in_R \mathbb{Z}_q$, for $j = 1, \ldots, t$.

2. He then broadcasts $\{\widetilde{R}_k = g_1^{r_k} g_2^{r_k'}\}_{k \in T}$ and sends $([s_i]^q, [s_i']^q) = (g(i), g'(i))$ to P_i.

3. Each P_i computes $\{\widetilde{S}_j = \prod_{k \in T} \widetilde{R}_k^{j^k}\}_{j \in I(\mathcal{P})}$ and verifies $([s_i]^q, [s_i']^q)$ as follows:

$$\widetilde{S}_i \stackrel{?}{=} g_1^{[s_i]^q} g_2^{[s_i']^q}$$

If the verification fails, the player broadcasts a complaint against the dealer. If there are more than t complaints, the dealer is disqualified. If there are less than t complaints, the dealer has to broadcast $([s_i]^q, [s_i']^q)$ for each complaining P_i. If the pair is correct, P_i is disqualified, otherwise the dealer.

Protocol PVSS2

Input (Dealer): $s \in M$

Output (P_i): $([s_i]^q, [s_i']^q)$, $\{\widetilde{S}_j\}_{j \in I(\mathcal{Q})}$

1. Analogous to step 1 of protocol PVSS.

2. The dealer computes $\{\widetilde{R}_k = g_1^{r_k} g_2^{r_k'}\}_{k \in T}$ and a non-interactive proof τ to show that $\widetilde{R}_0$ is a P-commitment of a secret which lies in M.

3. The dealer broadcasts $\{\tau, \widetilde{R}_k\}_{k \in T}$ and sends $([s_i]^q, [s_i']^q) = (g(i), g'(i))$ to P_i.

4. In addition to step 3 of protocol PVSS, each P_i verifies if $V(\tau, \widetilde{R}_0) = 1$. If the verification fails, then P_i sends a complaint against the dealer. If there are more than t complaints, the dealer is disqualified, otherwise the complaining players.

Fig. 5.4: Two Variants of Pedersen's Verifiable Secret Sharing Scheme.

available if t players have been disqualified. Although Feldman's protocol is based on poly-sharing, it only provides computational security because D-commitments are only computational hiding. This gives the following result (without proof):

Theorem 5.2.6 *Feldman's verifiable secret sharing scheme provides computational security if at most $t < n/2$ players are actively dishonest.*

Pedersen's Verifiable Secret Sharing: In [Ped91b], Pedersen modified Feldman's scheme, such that it provides unconditional security against $t < n/2$ active adversaries, replacing D-commitments with P-commitments. As shown in Section 5.2.2.2, such commitments are perfect hiding, which, together with the unconditionally security of poly-sharing, gives the following result (without proof):

Theorem 5.2.7 *Pedersen's verifiable secret sharing scheme provides unconditional security if at most $t < n/2$ players are actively dishonest.*

Pedersen's protocol, denoted by PVSS, is shown at the top of Figure 5.4. It works analogously to Feldman's approach, but requires some additional computations because the blinding in the P-commitment of the secret needs to be shared as well. Thus, the dealer has to choose two random polynomials $g(x)$ and $g'(x)$ and broadcast a set of P-commitments $\{\widetilde{R}_j = g_1^{r_j} g_2^{r'_j}\}_{j \in \mathcal{T}}$, where $r_0 := s$ and $r'_0 := s'$. Furthermore, he sends the shares $[s_i]^q = g(i)$ and $[s'_i]^q = g'(i)$ to player P_i who can verify their correctness straightforwardly. The remainder of the protocol and the reconstruction phase work analogously to Feldman's scheme.

Verifiable Sharing of a Secret that lies in a specific Interval: In Section 5.2.3.3, a protocol was described that can be used to prove that a secret, corresponding to a given P-commitment, is restricted to certain bit-positions. This approach can be combined with the basic variant of Pedersen's verifiable secret sharing scheme resulting in protocol PVSS2, shown at the bottom of Figure 5.4. In addition to the basic protocol, this extended version requires that the dealer runs the mentioned proof stated in Section 5.2.3.3 for the commitment $\widetilde{R}_0$.

5.2.4.3   Re-Sharing of Shared Secrets

Let $\mathcal{P}$ and $\mathcal{S}$ be two sets of players, where $|\mathcal{P}| = n$ and $|\mathcal{S}| = m$. Furthermore, let $s \in \mathbb{Z}_q$ be a secret that has been shared over $\mathcal{P}$ using a t-degree polynomial, i.e. each P_i holds a share $[s_i]^q$. The goal now is that s is shared over the players in $\mathcal{S}$, but without intermediate reconstruction. A protocol which gives a solution to this problem is called re-sharing or re-distribution protocol. A basic solution, which is widely used as a building block, can be sketched as follows [DJ97, WWW02]: each P_i shares $[s_i]^q$ over the players in $\mathcal{S}$ by using a t-degree sharing polynomial. Hence, each S_j receives the sub-share $[s_{ij}]^q$ from each player P_i. Then, S_j computes $\sum_{i=1}^{n}[s_{ij}]\lambda_i$ and gets a share $[s_j]^q$ of s. The correctness of the re-sharing procedure can be seen from the following:

$$ s = \sum_{j=1}^{m}[s_j]^q \lambda_j = \sum_{j=1}^{m}(\sum_{i=1}^{n}[s_{ij}]^q \lambda_i)\lambda_j = \sum_{i=1}^{n}(\sum_{j=1}^{m}[s_{ij}]^q \lambda_j)\lambda_i = \sum_{i=1}^{n}[s_i]^q \lambda_i $$

Notice that this technique can also be used to change the threshold.

5.2.5   Secure Multi-Party Computation

Sharing and reconstructing a secret is covered by secret sharing techniques. However, in several scenarios, secrets are used for computations (e.g. decryptions) or need to be modified from time to time (e.g. key updates). This of course is trivially possible if the secret is reconstructed immediately, but once this has been done, trust is no longer shared. Therefore techniques are required to either *use* or *modify* secrets without intermediate reconstruction.

The key to achieve this goal is *secure multi-party computation*. First mentioned by Yao in [Yao82a], research activities have led to practical solutions over the last two decades. Some remarkable results can be found in [GMW87, GV88, BOGW88, CCD88, Bea91, CDM00] and [HM00]. The idea behind secure multi-party computation is the following: there is a set $\mathcal{P}$ of n players, some of which know one or more secrets in $\mathbb{Z}_q$. These secrets are necessary for the secure evaluation of a function defined over $\mathbb{Z}_q$. Thereby, each player must not gain any information about the secret inputs of the other players. Thus, the idea is to share each secret over the other players allowing computations to be made using these shares exclusively. In the end, each player holds a share of the result, which can then be reconstructed or remains shared. According to [BOGW88], the main requirements for secure multi-party computation are the following:

R1 (Privacy): A player gets no information about secret inputs of other players.

R2 (Correctness): A player is not able to enforce incorrect results.

Over the years, several further requirements have been set (cf. [Lin03], for instance), such as *independence of inputs*, *robustness* (also known as *guaranteed output delivery*) and *fairness*. Special attention must be paid to robustness, which must hold if a computation is required to withstand active attacks. The requirement of fairness can be seen as a variant of privacy, since a player is required to not gain any information about intermediate results.

A secure multi-party computation generally consists of three stages [BOGW88]:

1. *Input-Stage:* Every player who inputs a secret shares it over $\mathcal{P}$.

2. *Computation-Stage:* The players perform computations using the shared secrets.

3. *Output-Stage:* The output is reconstructed using the corresponding shares.

Secret sharing techniques might be involved in all stages as a building block. Protocols must be provided for the computation-stage, such that shared computations result in the same output as computations which are performed directly on the secrets. In this work, solutions based on threshold secret sharing are used. Since computations are done in $\mathbb{Z}_q$, protocols to share modular addition and modular multiplication are necessary. Hereby, the following three basic protocols are of interest [BOGW88, GRR98]:

1. *Shared Addition of two Shared Secrets*: Given the shares of two secrets $a \in \mathbb{Z}_q$ and $b \in \mathbb{Z}_q$, the players want to compute $c = a + b$ without reconstruction of a or b.

2. *Shared Multiplication of a Shared Secret by a Public Constant*: Given a public constant $C \in \mathbb{Z}_q$ and the shares of a secret $a \in \mathbb{Z}_q$, the players want to compute $c = Ca$ without reconstruction of a.

3. *Shared Multiplication of two Shared Secrets*: Given the shares of two secrets $a \in \mathbb{Z}_q$ and $b \in \mathbb{Z}_q$, the players want to compute $c = ab$ without reconstruction of a or b.

In the following, instances of the above protocols are presented for the presence of passive adversaries (Section 5.2.5.1) and active adversaries (Section 5.2.5.2). Thereby, it is assumed that the adversaries behave staticly, which means that they decide at the beginning of the protocols if they are adversaries or not. Furthermore, we assume that a broadcast channel exists. In addition, results that achieve unconditional security with respect to the evaluation process require unconditionally secure encryption.

5.2.5.1   Basic Solutions with Security against Passive Adversaries

If only passive adversaries are expected, it suffices to guarantee that up to t players gain no useful information about the secrets, intermediate results or final results. Consider two secrets a and b, which have been shared among the set $\mathcal{P}$ during the input-stage by use of the following polynomials:

$$g_a(x) = a + \sum_{j=1}^{t} r_{a_j} x^j, \quad g_b(x) = b + \sum_{j=1}^{t} r_{b_j} x^j, \quad r_{a_j}, r_{b_j} \in_R \mathbb{Z}_q$$

Each player P_i possesses the shares $[a_i]^q = g_a(i)$ and $[b_i]^q = g_b(i)$.

Shared Addition of two Shared Secrets: The players want to compute $c = a + b$ using the shares of a and b exclusively. Adding two shares $[a_i]^q$ and $[b_i]^q$ gives

$$[a_i]^q + [b_i]^q = \left(a + \sum_{j=1}^{t} r_{a_j} i^j \right) + \left(b + \sum_{j=1}^{t} r_{b_j} i^j \right) = (a+b) + \sum_{j=1}^{t} \underbrace{(r_{a_j} + r_{b_j})}_{=: r_{c_j}} i^j$$

$$= c + \sum_{j=1}^{t} r_{c_j} i^j = [c_i]^q.$$

Hence, two shared secrets can be added using the shares without any communication. This holds because adding two t-degree polynomials results in a t-degree polynomial.

Shared Multiplication of a Shared Secret by a Public Constant: The players want to compute $c = Ca$, where $C \in \mathbb{Z}_q$ is public. Computing $C[a_i]^q$ gives

$$C[a_i]^q = C\left(a + \sum_{j=1}^{t} r_{a_j} i^j\right) = Ca + \sum_{j=1}^{t} \underbrace{Cr_{a_j}}_{=:r_{c_j}} i^j = c + \sum_{j=1}^{t} r_{c_j} i^j = [c_i]^q.$$

Again, no communication is necessary and the new share lies on a t-degree polynomial.

Naive Shared Multiplication of two Shared Secrets: Now consider the case where the players want to compute $c = ab$ just using the corresponding shares:

$$[a_i]^q [b_i]^q = \left(a + \sum_{j=1}^{t} r_{a_j} i^j\right)\left(b + \sum_{j=1}^{t} r_{b_j} i^j\right) = ab + \underbrace{(ar_{b_1} + r_{a_1}b)}_{r_{c_1}} i^1 + \ldots + \underbrace{r_{a_t} r_{b_t}}_{r_{c_{2t}}} i^{2t}$$

$$= c + \sum_{j=1}^{2t} r_{c_j} i^j = [c_i]^q$$

Hence, $[c_i]^q$ is a share of c, but unfortunately lies on a $2t$-degree polynomial. Hence, to reconstruct c, at least $2t+1$ shares are necessary. If more shared secrets are multiplied, then the resulting shares soon lie on a polynomial, whose degree exceeds the number of existing players. In such cases, a correct reconstruction or further computations are no longer possible. Hence, this straightforward solution is inflexible and inefficient.

Efficient Shared Multiplication of two Shared Secrets: In [GRR98], a solution was proposed which overcomes the drawbacks stated above. Every player P_i shares $[d_i]^q = [a_i]^q[b_i]^q$ (which lies on a $2t$-degree polynomial) over the other players by using a random t-degree polynomial. Then, each player locally performs an interpolation, resulting in a share $[c_i]^q$ of c which lies on a t-degree polynomial. It is obvious that for such a protocol at least $2t + 1$ players have to cooperate (remember that $[d_i]^q$ lies on a $2t$-degree polynomial). Hence, for correctness $t < n/2$ must hold. Notice that when multiplying several shared secrets, the players have to perform the intermediate re-sharing procedure *every time* two secrets have been multiplied, otherwise the degree increases rapidly. Now consider why the approach in [GRR98] is correct. Assume that every player P_i has already shared $[d_i]^q = [a_i]^q[b_i]^q$ over all other players by using a t-degree polynomial. Then, according to [GRR98], each player P_j locally computes $[c_j]^q = \sum_{i=1}^{n} [d_{ij}]^q \lambda_i$. Without loosing generality, consider the minimal number of necessary shares, i.e. $t + 1$ and $2t + 1$. Thus:

$$c = \sum_{j=1}^{2t+1} [d_j]^q \lambda_j = \sum_{j=1}^{2t+1} \left(\sum_{i=1}^{t+1} [d_{ji}]^q \lambda_i\right) \lambda_j = \sum_{i=1}^{t+1} \left(\sum_{j=1}^{2t+1} [d_{ji}]^q \lambda_j\right) \lambda_i = \sum_{i=1}^{t+1} [c_i]^q \lambda_i$$

It follows that $[c_i]^q$ is a valid share of c lying on a t-degree polynomial. Notice that the degree-reduction steps follow the principles of re-sharing as described in Section 5.2.4.3. The above considerations result in the following theorem (without proof).

Theorem 5.2.8 *A set of n players can evaluate any function in $\mathbb{Z}_q$ with unconditional security if at most $t < n/2$ of them are passively dishonest.*

5.2.5.2 Basic Solutions with Security against Active Adversaries

In the previous section, protocols based on standard poly-sharing according to Shamir were given. To make secret sharing secure against active adversaries, steps performed by the dealer or a share-owner need to be verifiable. The same mechanisms are now used to achieve security against active adversaries for secure multi-party computation.

If the homomorphic property of the used commitment scheme is not sufficient to verify particular protocol steps, it might be necessary to involve interactive proofs. In the following, verifiable computations based on P-commitments are considered, since the subsequent protocols are also based on Pedersen's verifiable secret sharing. Again, two secrets a and b are considered, which have been shared among the set $\mathcal{P}$ during the input-stage, but this time using PVSS. Hence, each P_i holds the shares $[a_i]^q$, $[a_i']^q$, $[b_i]^q$ and $[b_i']^q$, and the set $\{\widetilde{A}_j, \widetilde{B}_j\}_{j \in I(\mathcal{Q})}$.

Shared Verifiable Addition of two Shared Secrets: The players want to compute shares of $c = a + b$ and update the corresponding set of commitments. Each P_i computes $[c_i]^q = [a_i]^q + [b_i]^q$, $[c_i']^q = [a_i']^q + [b_i']^q$ and defines the corresponding set of commitments $\{\widetilde{C}_j = \widetilde{A}_j \widetilde{B}_j\}_{j \in I(\mathcal{Q})}$. Notice that this protocol needs no interaction between the players. The modified set of commitments is necessary for later use.

Shared Verifiable Multiplication of a Shared Secret by a Public Constant: For a constant $C \in \mathbb{Z}_q$, each P_i computes $[c_i]^q = C[a_i]^q$, $[c_i']^q = C[a_i']^q$ and forms the corresponding set $\{\widetilde{C}_j = \widetilde{A}_j^C\}_{j \in I(\mathcal{Q})}$. Again no communication is required.

Shared Verifiable Multiplication of two Shared Secrets: For the multiplication protocol some intrinsic modifications are necessary. The problem is that the homomorphic property of P-commitments is not sufficient to make the protocol steps of the basic multiplication protocol verifiable, i.e. given $\widetilde{A}_j$ and $\widetilde{B}_j$, computing $\widetilde{R}_{i0}$ is infeasible if the Diffie-Hellman Problem is hard in $\mathbb{G}_q$. Hence, when sharing $[d_i]^q = [a_i]^q[b_i]^q$ by use of PVSS, P_i has to prove that $(\widetilde{A}_i, \widetilde{B}_i, \widetilde{R}_{i0})$ is a Pedersen-Diffie-Hellman-Triple, which can be done using the protocol given in Section 5.2.3.2. In the context of multi-party computation, a Σ-proof can be basically realized in the following three ways:

1. P_i runs a separate proof with all other players.

2. P_i runs one proof, where the other players act as a shared verifier.

3. P_i broadcasts a non-interactive version of the proof.

Protocol VMUL

Input (P_i): $([a_i]^q, [a_i']^q)$, $([b_i]^q, [b_i']^q)$, $\{\widetilde{A}_j, \widetilde{B}_j\}_{j \in I(\mathcal{Q})}$

Output (P_i): $([c_i]^q, [c_i']^q)$, $\{\widetilde{C}_j\}_{j \in I(\mathcal{Q})}$

1. Each P_i computes $[d_i]^q = [a_i]^q [b_i]^q$ and $[d_i']^q = [a_i']^q [b_i']^q$.

2. Then each P_i runs the following special variant of PVSS to share $[d_i]^q$:

   a) Analogous to step 1 of protocol PVSS.

   b) P_i computes $\{\widetilde{R}_{ik} = g_1^{r_{ik}} g_2^{r_{ik}'}\}_{k \in \mathcal{T}}$ and generates a non-interactive proof τ_i to show that $(\widetilde{A}_i, \widetilde{B}_i, \widetilde{R}_{i0})$ is a Pedersen-Diffie-Hellman-Triple.

   c) P_i broadcasts $\{\tau_i, \widetilde{R}_{ik}\}_{k \in \mathcal{T}}$ and sends $([d_{ij}]^q, [d_{ij}']^q) = (g_i(j), g_i'(j))$ to P_j.

   d) In addition to step 3 of protocol PVSS, each P_j verifies if $V(\tau_i, (\widetilde{A}_i, \widetilde{B}_i, \widetilde{R}_{i0})) = 1$ for every i. If for any i the verification fails, then P_j sends a complaint against P_i. If there are more than t complaints, then P_i is disqualified.

3. Finally, each P_j computes $[c_j]^q = \sum_{i \in I(\mathcal{Q})} [d_{ij}]^q \lambda_i$, $[c_j']^q = \sum_{i \in I(\mathcal{Q})} [d_{ij}']^q \lambda_i$ and the set of P-commitments of all other shares: $\{\widetilde{C}_k = \prod_{i \in I(\mathcal{Q})} \widetilde{D}_{ik}^{\lambda_i}\}_{k \in I(\mathcal{Q})}$.

Fig. 5.5: Shared Verifiable Multiplication of two Shared Secrets based on P-commitments.

The first variant is obviously quite inefficient. In the second variant, the prover broadcasts the first-message. Then, all other players jointly compute a challenge and send their shares to the prover who reconstructs it. Then he broadcasts the response and every player can locally verify the proof. This solution is also quite inefficient. The third variant is practical, since no additional (regarding PVSS) communication is necessary. Moreover, Σ-proofs are only honest-verifier zero-knowledge, but if made non-interactive, they can be even used in the presence of dishonest verifiers.

At the end of protocol VMUL, each player computes his share of $c = ab$ and of $c' = a'b'$. An output of PVSS, that each player receives (cf. step 3 of PVSS), is the set $\{\widetilde{D}_{ik}\}_{i,k \in I(\mathcal{Q})}$ containing the P-commitment of each sub-share $[d_{ik}]^q$. These commitments are sufficient to compute the set $\{\widetilde{C}_k\}_{k \in I(\mathcal{Q})}$. The above considerations result in the following theorem (without proof):

Theorem 5.2.9 *A set of n players can evaluate any function in $\mathbb{Z}_q$ with computational security if at most $t < n/2$ of them are actively dishonest.*

5.2.6 Discrete-Logarithm-Based Threshold Cryptosystems

In general, threshold cryptosystems are public key cryptosystems where the secret key only exists in shared form and computations, such as decryption or signature generation, are done over the corresponding shares exclusively [Des88, DF90]. On the contrary, the public key remains in reconstructed form, so that it can be used in the non-threshold setting. Several threshold cryptosystems have been proposed since [Des88, DF90]. A good survey (at least covering up until 1997) can be found in [Des98]. Basic distributed key generation protocols for discrete-logarithm-based schemes have been suggested in [Ped91a, GJKR99] and for RSA-like approaches in [GJKR96, Coc97, FMY98, BF01] and [ACS02]. In the context of threshold decryption, some results can be found in [SG98, CG99, CGJ+99] and [FPS01]. For threshold signature generation the interested reader might have a look at the approaches in [DF92, Har94, GJKR96, Sho00, DD05, JS05] and [ADN06]. Apart from key generation, decryption and signature generation, there exist several other aspects in public-key cryptography where distributed computations take place. Proactive schemes, for instance, include key updates, updates of shares and re-encryptions. Some corresponding results can be found in [OY91, HJKY95, FGMY97] and [Rab98], respectively.

In the following, the basic solutions of [Ped91a, GJKR99] for discrete-logarithm-based distributed key generation are presented. These solutions are a building block for the protocols presented later on in this chapter. For a simple illustration of threshold decryption and threshold signature generation the scheme proposed in [DF90] is examined, and an additional threshold version of an ElGamal-variant proposed in [HPM94] is presented. For the sake of generality, the presented key generation protocols output shares of a secret z and the corresponding public key $Z = g^z$. In the presented ElGamal threshold decryption and signature protocols the keys are denoted as usually done throughout this work (cf. Section 2.2), i.e. z and $h = g^z$ for ElGamal encryption and x and $y = g^x$ for ElGamal signature generation.

5.2.6.1 Distributed Key Generation

In [Ped91a], Pedersen described the first distributed key generation protocol for discrete-logarithm-based cryptosystems. The basic idea in [Ped91a] is to define the secret key $z \in \mathbb{Z}_q$ as the modular sum of random values, chosen by a set of players. For obvious reasons, no player is allowed to know anything about the input of any other player. Hence, it is necessary to use secure multi-party computation to compute the following:

$$z = \sum_{i=1}^{n} z_i, \quad z_i \in_R \mathbb{Z}_q$$

where z_i is the secret input of player P_i. During the input-stage, every player P_i has to share z_i among the players of $\mathcal{P}$ using polynomial sharing. So, every P_j receives the poly-shares $[z_{1j}]^q, \ldots, [z_{nj}]^q$. During the computation-stage, each player locally adds the received shares resulting in a share of the sum z. Since the goal is *not* to reconstruct

z, this multi-party computation has no output-stage. The basic construction needs to be extended, such that the D-commitment $Z = g^z$ (the public key) can be efficiently computed. In [Ped91a], Feldman's verifiable secret sharing was used so that every player gets Z as a by-product. Moreover, security against $t < n/2$ active adversaries can be achieved using verifiable secret sharing. As a consequence z is the sum of the secrets submitted by the *honest* players exclusively.

In [GJKR99], it has been shown that Pedersen's protocol is insecure against an attack where an adversary forces the protocol to output a non-uniformly distributed secret key. This is possible because D-commitments are used, which are only computational hiding. In [GJKR03], the authors showed that Pedersen's protocol can still be used for some schemes, such as a threshold variant of Schnorr's signature scheme. In [GJKR99, GJKR07], solutions have been given that overcome the problem of Pedersen's approach. The main idea is to replace D-commitments with P-commitments and adapt the protocol appropriately. As mentioned, in Pedersen's key generation protocol each player gets $Z = g^z$ as a by-product. If, however, P-commitments are used, then each player obtains $\widetilde{Z} = g_1^z g_2^{z'}$. Additional steps are necessary to obtain $Z = g^z$ and to guarantee that the committed secrets are equal.

In [GJKR99], the following minimal requirements for secure distributed key generation for discrete-logarithm-based cryptosystems were set:

R1 (Correctness): Here the following four cases are distinguished:

(a) Any $t + 1$ shares provided by honest players define the same secret $z \in \mathbb{Z}_q$.
(b) Correct shares of z can be distinguished from incorrect ones.
(c) All honest parties have the same public key Z of the corresponding secret key z.
(d) The secret key z is uniformly distributed over $\mathbb{Z}_q$.

R2 (Secrecy): No information on z can be learned except for what is implied by Z.

For the sake of simplicity, a protocol where every player chooses a random value and then a multi-party addition is performed, is sometimes called *joint random (verifiable) secret sharing*. In [GJKR99], Pedersen's verifiable secret sharing is used and hence the corresponding joint random verifiable secret sharing is denoted as JPVSS. The protocol can be found in Figure 5.6 and works as follows: every player P_i generates a random secret $z_i \in \mathbb{Z}_q$ and shares it over the other players using PVSS. Steps 2-3 represent several implicit runs of the addition protocol sketched in Section 5.2.5.2.

The distributed key generation protocol DKG of [GJKR99] can now be sketched as follows: the players jointly generate shares of a random secret z using JPVSS. Then, every P_i broadcasts D-commitments of each of his sharing-coefficient r_{ik} and every receiver P_j verifies if they correspond to the share $[z_{ij}]^q$ which he received from P_i in the

---

Protocol JPVSS

Output (P_i): $([z_i]^q, [z_i']^q)$, $\{[z_{ji}]^q, [z_{ji}']^q, \widetilde{Z}_j\}_{j \in I(\mathcal{Q})}$, $\{r_{ik}\}_{k \in \mathcal{T}}$

1. Each player P_i chooses $z_i \in \mathbb{Z}_q$ and shares it among the other players using PVSS. Hence, P_j gets $\{([z_{ij}]^q, [z_{ij}']^q)\}_{i \in I(\mathcal{Q})}$ and $\{\widetilde{Z}_{ik}\}_{i,k \in I(\mathcal{Q})}$ as output and stores the set of his own sharing coefficients $\{r_{jk}\}_{k \in \mathcal{T}}$.

2. Now each player P_j computes $[z_j]^q = \sum_{i \in I(\mathcal{Q})} [z_{ij}]^q$ and $[z_j']^q = \sum_{i \in I(\mathcal{Q})} [z_{ij}']^q$.

3. Finally, each player computes the set of P-commitments $\{\widetilde{Z}_j = \prod_{i \in I(\mathcal{Q})} \widetilde{Z}_{ij}\}_{j \in I(\mathcal{Q})}$.

Protocol DKG

Output (P_i): $[z_i]^q$, Z

1. Each player P_i participates in a run of JPVSS. Hereby, he obtains $([z_i]^q, [z_i']^q)$, $\{([z_{ji}]^q, [z_{ji}']^q), \widetilde{Z}_j\}_{j \in I(\mathcal{Q})}$ and the set of his own sharing coefficients $\{r_{ik}\}_{k \in \mathcal{T}}$ that correspond to his secret input z_i in JPVSS.

2. Then, each P_i broadcasts $\{R_{ik} = g^{r_{ik}}\}_{k \in \mathcal{T}}$.

3. Each P_j computes $\{Z_{ij} = g^{[z_{ij}]^q}\}_{i \in I(\mathcal{Q})}$ and verifies if $Z_{ij} = \prod_{k \in \mathcal{T}} R_{ik}^{j^k}$ holds for all $i \in I(\mathcal{Q})$. If the check fails for any i, P_j complains against P_i by broadcasting $([z_{ij}]^q, [z_{ij}']^q)$, which satisfies the verification in PVSS but not the one above. If the complaint is valid, the remaining players jointly reconstruct z_i.

4. Finally, each player computes the public key $Z = \prod_{j \in I(\mathcal{Q})} R_{j0}$.

Fig. 5.6: Distributed Key Generation.

run of protocol JPVSS. If the verification fails, P_j complains against P_i by broadcasting $([z_{ij}]^q, [z_{ij}']^q)$. For all valid complaints, the honest players have to reconstruct z_i. In step 4, each player computes Z using the D-commitments of the honest players, as broadcasted in step 2.

5.2.6.2 ElGamal Threshold Decryption

In [DF90], the first threshold decryption scheme was proposed. The basic principle of such a scheme is the following: once the secret key exists in shared form, the decryption function can be efficiently performed using the ciphertext resulting in the plaintext. An intrinsic property is that during the decryption, no information about the secret key can be obtained. In [DF90], a solution is given for the ElGamal decryption algorithm, however, only with security against passive adversaries. Here, this basic solution is sketched and then extended using the techniques presented in [CP93, CGS97] to achieve security against active adversaries (without proofs).

During the ElGamal encryption in $\mathbb{G}_q$, a plaintext $m \in \mathbb{G}_q$ is encrypted using the public

---

Protocol ElGamal-TDEC

Input (P_i): $[z_i]^q$, (u, e), $\{h_j\}_{j \in I(\mathcal{Q})}$

Output (P_i): m

1. Each player P_i computes $u_i = u^{-[z_i]^q}$.

2. Then he computes a non-interactive proof τ_i to ensure that $dlog_g(h_i) = dlog_u(u_i^{-1})$ and broadcasts the pair (τ_i, u_i) to the players in $\mathcal{P}$.

3. For each i every P_j checks if $V(\tau_i, (h_i, u, u_i)) = 1$. If the check fails, he broadcasts a complaint against P_i. If there are more than t complaints, then P_i is disqualified.

4. Each player computes $m = e \prod_{j \in I(\mathcal{Q})} u_j^{\lambda_j}$ over the values of the honest players.

Fig. 5.7: ElGamal Threshold Decryption.

key $h = g^z$ and a randomizer $r \in_R \mathbb{Z}_q$, resulting in the ciphertext $(u, e) = (g^r, mh^r)$. Given z and (u, e), the plaintext can be obtained by computing $m = eu^{-z}$. Now, assume that z is shared among a set of n players, i.e. each P_i holds a share $[z_i]^q$. Then, by using Lagrange interpolation, the decryption can be written as follows:

$$eu^{-z} = eu^{-\sum_{i=1}^{n} [z_i]^q \lambda_i} = e \prod_{i=1}^{n} u^{-[z_i]^q \lambda_i} = e \prod_{i=1}^{n} \underbrace{(u^{-[z_i]^q})}_{=: u_i}{}^{\lambda_i} = e \prod_{i=1}^{n} u_i^{\lambda_i}$$

Every player P_i can compute $u_i = u^{-[z_i]^q}$ independently. Moreover, if the player broadcasts u_i, the scheme still remains secure under the assumption that computing discrete logarithms is hard. Given $u_1, \ldots, u_n$ and e, every player computes m.

If any player is actively dishonest, then the decryption results in an invalid plaintext. Unfortunately, in the above sketched protocol, none of the players can verify if u_i is correct. This drawback can be overcome using an interactive proof for the equality of discrete logarithms [CP93]. Using the received commitments in DKG for the shared generation of z, each player can compute the D-commitment h_i corresponding to $[z_i]^q$. So, every P_i just has to prove that $dlog_g(h_i) = dlog_u(u_i^{-1})$ holds. Such a proof can be made non-interactive: P_i computes $c_i = \mathcal{H}(g^{\alpha_i} || u^{\alpha_i})$, where $\alpha_i \in \mathbb{Z}_q$, and $s_i = \alpha_i - c_i[z_i]^q$, before setting $\tau_i := (c_i, s_i)$. Given τ_i and the triple (h_i, u, u_i), any player can verify τ_i with the function $V : T \times \mathbb{G}_q^3 \to \{0, 1\}$, which is defined as follows:

$$V(\tau, (h_i, u, u_i)) := \begin{cases} 0 & \text{if } \tau_{i1} \neq \mathcal{H}(h_i^{\tau_{i1}} g^{\tau_{i2}} || u_i^{-\tau_{i1}} u^{\tau_{i2}}) \\ 1 & \text{otherwise} \end{cases}$$

It can be shown that the above proof is a Σ-proof (analogously to Section 5.2.3.2).

5.2.6.3 ElGamal Threshold Signature Generation

Using a threshold signature scheme, a valid signature can be generated if at least $t+1$ players agree to collaborate. It seems as if a solution can be designed analogously to threshold decryption. A threshold version of the ElGamal signature scheme requires similar protocol steps as used for the threshold decryption. Additionally, the players have to jointly generate shares of a randomizer.

In [HPM94], several variants of the ElGamal signature scheme have been presented. Here, one of these approaches is used for $\mathbb{G}_q < \mathbb{Z}_p^*$, where $q|(p-1)$ and p, q are both odd primes. The signature $(R, s) \in \mathbb{G}_q \times \mathbb{Z}_q$ of a message $m \in \mathbb{Z}_q$ is computed as follows: set $R = g^r$ and $s = xR + rm$, where $r \in_R \mathbb{Z}_q$ and $x \in \mathbb{Z}_q$ is the signature generation key. The signature (R, s) is accepted to be valid if $g^s = y^R R^m$ holds.

Protocol ElGamal-TSIG

Input (P_i): $[x_i]^q$, m, $\{y_j\}_{j \in I(\mathcal{Q})}$

Output (P_i): (R, s)

1. The players participate in a run of DKG to generate shares of $r \in \mathbb{Z}_q$. Hence, each P_i gets the output $[r_i]^q$ and the D-commitment $R = g^r$.

2. Each player P_i broadcasts $[s_i]^q = [x_i]^q R + [r_i]^q m$.

3. For each i every P_i checks if $g^{[s_i]^q} = y_i^R R_i^m$. If the check fails, he broadcasts a complaint against P_i. If there are more than t complaints, then P_i is disqualified.

4. Each player computes $s = \sum_{i \in I(\mathcal{Q})} [s_i]^q \lambda_i$ and forms the signature (R, s).

Fig. 5.8: ElGamal Threshold Signature Generation.

Now consider the threshold setting: each player P_i is provided with a poly-share $[x_i]^q$ of $x \in \mathbb{Z}_q$ and the set $\{y_j = g^{[x_j]^q}\}_{j \in I(\mathcal{Q})}$. If the players use DKG to generate shares of r, then each player P_i is provided with the commitment $R = g^r$ as a by-product. Moreover, each player can locally compute a D-commitment R_i of each share $[r_i]^q$. However, s must be generated without revealing any information about r and x. Consider $s = xR + rm$, where x and r are represented by their Lagrange representation:

$$s = \left(\sum_{i \in I(\mathcal{Q})} [x_i]^q \lambda_i \right) R + \left(\sum_{i \in I(\mathcal{Q})} [r_i]^q \lambda_i \right) m = \sum_{i \in I(\mathcal{Q})} \underbrace{([x_i]^q R + [r_i]^q m)}_{=: [s_i]^q} \lambda_i = \sum_{i \in I(\mathcal{Q})} [s_i]^q \lambda_i$$

Since R and m are public values, the above formula can be computed using secure multi-party addition. All that remains is to perform the reconstruction phase of PVSS to compute s and then output the signature (R, s).

5.3   Fusion-Based Threshold Schemes

In the following, protocols for secure multi-party computation in $\mathbb{F}_p$ are given. It is shown that the protocols have *the same communication complexity* and the *same level of security* as the standard solutions for computations in $\mathbb{Z}_q$.

5.3.1   Secret Sharing

When designing a secret sharing scheme, one main goal is to keep the communication complexity as low as possible. If done straight forwardly, sharing a secret $s \in \mathbb{F}_p$ could be done using some kind of "fusion-polynomial" defined as follows:

$$g(\mathsf{x}) = \mathsf{s} + \mathsf{r}_1 \mathsf{x} + \ldots + \mathsf{r}_t \mathsf{x}^t, \quad \mathsf{r}_i \in_R \mathbb{F}_p \qquad (5.3.1)$$

Notice that $\mathsf{x} = (x_1, x_2)$ and $(x_1, x_2)^j = (x_1, x_2) \cdot \ldots \cdot (x_1, x_2)$ (j times). As a consequence, each component of $g(\mathsf{x})$ is a t-degree bivariate polynomial. For instance, consider a secret $\mathsf{s} \in \mathbb{F}_p$ which has to be shared using a 2-degree fusion-polynomial $g(\mathsf{x})$. Let $\mathsf{s} = (a, b)$, $\mathsf{r}_1 = (c, d)$ and $\mathsf{r}_2 = (e, f)$, then:

$$
\begin{aligned}
g(\mathsf{x}) &= \mathsf{s} + \mathsf{r}_1 \mathsf{x} + \mathsf{r}_2 \mathsf{x}^2 \\
&= (a, b) + (c, d)(x_1, x_2) + (e, f)(x_1, x_2)^2 \\
&= (a, b) + (cx_1 - dx_2, cx_2 + dx_1) + (e, f)(x_1^2 - x_2^2, 2x_1 x_2) \\
&= (a + cx_1 - dx_2, b + cx_2 + dx_1) + (ex_1^2 - ex_2^2 - 2fx_1 x_2, 2ex_1 x_2 + fx_1^2 - fx_2^2) \\
&= (a + cx_1 - dx_2 + ex_1^2 - ex_2^2 - 2fx_1 x_2, b + cx_2 + dx_1 + 2ex_1 x_2 + fx_1^2 - fx_2^2)
\end{aligned}
$$

It can be seen that each component is a quadratic bivariate polynomial. For the interpolation of a t-degree bivariate polynomial, each player generally needs to be provided with at least $t + 1$ shares instead of one share if a t-degree univariate polynomial is used. Although this approach would work in general, it is quite inefficient and requires unnecessarily much space and computations.

A more efficient idea is to share *each component* of s using standard poly-sharing in $\mathbb{Z}_q$. A simple trick allows such sharing to be expressed using operations in $\mathbb{F}_p$ exclusively. One can prove by induction that $(x, 0)^j = (x^j, 0)$. Setting $\mathsf{x} = (x, 0)$ in (5.3.1) gives

$$
\begin{aligned}
g((x, 0)) &= \mathsf{s} + \mathsf{r}_1 (x, 0) + \ldots + \mathsf{r}_t (x, 0)^t \\
&= (s_1, s_2) + (r_{11}, r_{12})(x, 0) + \ldots + (r_{t1}, r_{t2})(x^t, 0) \\
&= (s_1, s_2) + (r_{11}x, r_{12}x) + \ldots + (r_{t1}x^t, r_{t2}x^t) \\
&= (s_1 + r_{11}x + \ldots + r_{t1}x^t, s_2 + r_{12}x + \ldots + r_{t2}x^t) = (g_1(x), g_2(x))
\end{aligned}
$$

The components $g_1(x)$ and $g_2(x)$ are t-degree sharing-polynomials in $\mathbb{Z}_q$. Hence, one can set $h(x) := g((x,0))$ and define the i-th fusion-share $\langle s_i \rangle^q$ as follows:

$$\langle s_i \rangle^q := h(i) = g((i,0)) = (g_1(i), g_2(i)) = ([s_{i1}]^q, [s_{i2}]^q)$$

Using the notation $\langle \cdot \rangle^q$ helps to distinguish fusion-shares from ordinary poly-shares denoted as $[\cdot]^q$. The reconstruction of s can be done by component-wise Lagrange interpolation in $\mathbb{Z}_q$, which can be expressed using $+$ and $\cdot$ as follows:

$$\begin{aligned}
s &= \langle s_1 \rangle^q (\lambda_1, 0) + \ldots + \langle s_n \rangle^q (\lambda_n, 0) \\
&= ([s_{11}]^q, [s_{12}]^q)(\lambda_1, 0) + \ldots + ([s_{n1}]^q, [s_{n2}]^q)(\lambda_n, 0) \\
&= ([s_{11}]^q \lambda_1, [s_{12}]^q \lambda_1) + \ldots + ([s_{n1}]^q \lambda_n, [s_{n2}]^q \lambda_n) \\
&= \Big(\sum_{i=1}^{n} [s_{i1}]^q \lambda_i, \sum_{i=1}^{n} [s_{i2}]^q \lambda_i \Big) = (s_1, s_2)
\end{aligned}$$

The following corollary naturally results from Theorem 5.2.5:

Corollary 5.3.1 *Shamir's secret sharing scheme with respect to $\mathbb{F}_p$ provides unconditional security if at most $t < n$ players are passively dishonest.*

Proof. From Theorem 5.2.5, poly-sharing in $\mathbb{Z}_q$ provides unconditional security against $t < n$ passive adversaries. In the fusion-variant of Shamir's scheme, ordinary poly-sharing in $\mathbb{Z}_q$ is applied component-wise using independent random polynomials. Hence, fusion-poly-sharing is unconditionally secure against $t < n$ passive adversaries. □

It must be shown that security against active adversaries can be achieved. In the following, the solutions in [Fel87] and [Ped91b] are shown to be feasible with respect to $\mathbb{F}_p$ and $\mathbb{G}_p$. Henceforth, let g, g_1 and g_2 be generating bases for $\mathbb{G}_p$ where the mutual fusion discrete logarithms are unknown.

5.3.1.1 Fusion Feldman Verifiable Secret Sharing

Feldman's verifiable secret sharing is based on D-commitments which are perfect binding and computational hiding. To realize Feldman's scheme with respect to fusion-poly-sharing and $\mathbb{G}_p$, D-commitments need to be defined appropriately:

Definition 5.3.1 Let $s \in \mathbb{F}_p$ be a secret. The *Fusion Discrete Logarithm Commitment* (FD-commitment) is given by $S = g^s$.

Lemma 5.3.1 *FD-commitments are (1) perfect binding and (2) computational hiding.*

Proof. (1) From Theorem 4.3.3 it directly follows that for every element $s \in \mathbb{F}_p$ exactly one corresponding element $S \in \mathbb{G}_p$ exists, such that $S = g^x$ holds. (2) Is equivalent to solving the Fusion Discrete Logarithm Problem. □

Let $\langle s_1 \rangle^q, \ldots, \langle s_n \rangle^q$ be the fusion-shares of a secret $s \in \mathbb{F}_p$ and $r_1, \ldots, r_t$ the corresponding fusion-coefficients. The FD-commitment S_i of each share $\langle s_i \rangle^q$ is

$$S_i = g^{\langle s_i \rangle^q} = g^{s + r_1(i,0) + \cdots + r_t(i,0)^t} = g^s (g^{r_1})^{(i,0)} \cdots (g^{r_t})^{(i^t,0)}$$

One can see, that Feldman's verifiable secret sharing scheme can be given analogously to Section 5.2.4.2. From Theorem 5.2.6 and Corollary 5.3.1 the following can be drawn:

Corollary 5.3.2 *Feldman's verifiable secret sharing scheme with respect to $\mathbb{F}_p$ provides computational security if at most $t < n/2$ players are actively dishonest.*

Proof. Poly-sharing in $\mathbb{F}_p$ provides unconditional security against $t < n$ passive adversaries. Since FD-commitments are computational hiding, the security is at most computational. The possible occurrence of active adversaries decreases the threshold to $t < n/2$ as usual.   □

5.3.1.2 Fusion Pedersen Verifiable Secret Sharing

To realize protocol PVSS, P-commitments must be defined appropriately:

Definition 5.3.2 Let $s \in \mathbb{F}_p$ be a secret. The *Fusion Pedersen Commitment* (FP-commitment) is given by $\tilde{S} = g_1^s g_2^{s'}$, where $s' \in_R \mathbb{F}_p$.

Lemma 5.3.2 *FP-commitments are (1) computational binding and (2) perfect hiding.*

Proof. Directly follows from Lemmata 5.2.2 and 5.3.1.   □

A protocol analogous to PVSS can be given straightforwardly. Henceforth, the fusion-variant of PVSS is denoted by FPVSS so that we renounce to give further details.

Corollary 5.3.3 *Pedersen's verifiable secret scheme with respect to $\mathbb{F}_p$ provides unconditional security if at most $t < n/2$ players are actively dishonest.*

Proof. This time FP-commitments are used. Since FP-commitments are perfect hiding, unconditional security against $t < n/2$ active adversaries holds.   □

5.3.2 Secure Multi-Party Computation

Surprisingly, it can be shown that if secret sharing in $\mathbb{F}_p$ is used, then secure multi-party computation in $\mathbb{F}_p$ has the same communication complexity as ordinary multi-party computation in $\mathbb{Z}_q$. In the following, the principles of adding two shared secrets, multiplying a shared secret by a public constant and multiplying two shared secrets are sketched. Therefore, let $a, b \in \mathbb{F}_p$ be two secrets that have been shared over n players using secret sharing in $\mathbb{F}_p$.

Shared Addition of two Shared Secrets: The goal is to compute $c = a + b$ over the corresponding shares. Computing $\langle a_i \rangle^q + \langle b_i \rangle^q$ gives

$$\langle a_i \rangle^q + \langle b_i \rangle^q = ([a_{i1}]^q, [a_{i2}]^q) + ([b_{i1}]^q, [b_{i2}]^q) = ([a_{i1}]^q + [b_{i1}]^q, [a_{i2}]^q + [b_{i2}]^q)$$

Hence, shared addition in $\mathbb{F}_p$ can be simulated over two runs of shared addition in $\mathbb{Z}_q$. Thus, no communication among the players is required.

Shared Multiplication of a Shared Secret by a Public Constant: The goal is to compute $c = Ca$, where $C \in \mathbb{F}_p$ is a public constant, using just C and the shares of a. Let $C = (C_1, C_2)$, then computing $C\langle a_i \rangle^q$ gives

$$C\langle a_i \rangle^q = (C_1, C_2)([a_{i1}]^q, [a_{i2}]^q) = (C_1[a_{i1}]^q - C_2[a_{i2}]^q, C_1[a_{i2}]^q + C_2[a_{i1}]^q)$$

Notice that both components are computed using shared addition and the shared multiplication by a public constant in $\mathbb{Z}_q$. Hence, no communication is necessary.

Efficient Shared Multiplication of two Shared Secrets: The goal is to compute $c = ab$ using the corresponding shares. Computing $\langle a_i \rangle^q \langle b_i \rangle^q$ gives

$$\langle a_i \rangle^q \langle b_i \rangle^q = ([a_{i1}]^q, [a_{i2}]^q)([b_{i1}]^q, [b_{i2}]^q) = ([a_{i1}]^q[b_{i1}]^q - [a_{i2}]^q[b_{i2}]^q, [a_{i1}]^q[b_{i2}]^q + [a_{i2}]^q[b_{i1}]^q)$$

One can see, that four runs of shared multiplication and two runs of shared addition in $\mathbb{Z}_q$ are necessary to evaluate $\cdot$ over two fusion-shared secrets. Shared addition requires no communication. Since the four multiplications are mutually independent, they can be run in parallel and hence no *additional* communication is necessary.

Of course multi-party computation in $\mathbb{F}_p$ requires more computations compared to schemes that are based on $\mathbb{Z}_q$. However, in shared cryptosystems it is much more important to minimize the number of sent messages. Fortunately, sharing $+$ and $\cdot$ in $\mathbb{F}_p$ has the same communication complexity as sharing $+$ and $\cdot$ in $\mathbb{Z}_q$. Notice that by expressing fusion-based multi-party computation using ordinary multi-party computation, there is no necessity to give extra security proofs. This leads to the following corollary (without proof).

Corollary 5.3.4 *A set of n players can evaluate any function in $\mathbb{F}_p$ with unconditional security if at most $t < n/2$ of them are passively dishonest.*

The above sketched protocols can be made verifiable using FD-commitments or FP-commitments together with the corresponding verifiable fusion secret sharing schemes. For shared addition and shared multiplication by a public constant, the homomorphic property of the commitment schemes is sufficient. For the shared multiplication protocol an interactive proof for the equality of fusion discrete logarithms is necessary. Such a proof can be given analogously to the ordinary setting, since the necessary properties hold through Theorem 4.3.2.

Corollary 5.3.5 *A set of n players can evaluate any function in $\mathbb{F}_p$ with computational security if at most $t < n/2$ of them are actively dishonest.*

5.3.3 Threshold Cryptosystems

Based on the techniques shown in the previous sections, threshold cryptosystems can be also realized in $\mathbb{G}_p$. Notice that the necessary sub-protocols need to be adapted in terms of computations, but no extra communication is necessary. For instance, consider DKG, where JPVSS is used as a building block. This protocol in turn requires one run of PVSS performed by the players and some broadcasts. Realizing DKG for $\mathbb{G}_p$ only requires that four random elements are shared in FPVSS instead of two. This can be done in parallel and hence requires no extra communication. Since the protocols can be designed straightforwardly for $\mathbb{G}_p$, the details are omitted.

5.4 Collision-Free Distributed Key Generation

In this section, the requirements for distributed key generation, as stated in Section 5.2.6.1, are adapted to take system-wide uniqueness into account. Furthermore, three concepts of collision-free distributed key generation for discrete-logarithm-based threshold cryptosystems are sketched on an abstract level. These concepts constitute the basis for the practical approaches given in the Sections 5.5, 5.6, 5.7 and 5.8. Finally, some common assumptions are given for these practical approaches.

5.4.1 Adapted Requirements

In Section 5.2.6.1, a secret key z is required to be uniformly distributed over $\mathbb{Z}_q$ (cf. R1(d)). In the current chapter, the goal is to additionally ensure that z is system-wide unique. This obviously always contradicts the requirement for uniform distribution. Notice, however, that R1(d) is just a necessary condition to fulfill the requirement for secrecy to the highest possible degree, i.e. the equivalence of finding z and the Discrete Logarithm Problem in groups of order q. Thus, we suggest to relax the requirement for correctness by removing R1(d), while keeping in mind that secrecy needs to be fulfilled sufficiently regarding the adversary model. Moreover, the requirement of uniqueness is added. Together this gives the following new list.

R1 (Correctness): Here the following three cases are distinguished:

(a) Any $t + 1$ shares provided by honest players define the same secret $z \in \mathbb{Z}_q$.
(b) Correct shares of z can be distinguished from incorrect ones.
(c) All honest parties have the same public key Z for the corresponding secret key z.

R2 (Secrecy): With respect to a defined adversary model, no advantageous information on z can be learned except for what is implied by Z.

R3 (Uniqueness): The secret key z is system-wide unique.

Notice that the requirement for secrecy is relaxed, such that one might be able to obtain *some* information on the secret, but not enough to break the subsequent cryptosystem. Additionally, it is hardly possible to achieve uniform distribution as required in Section 5.2.6.1 because this implies the existence of true random number generators.

For fusion-based cryptosystems, the requirements can be formulated analogously. Thus, their additional formulation is not given here for simplicity.

5.4.2 Shared Collision-Free Number Generation

Recall the core idea in the design of CFNG1 (cf. Section 3.4). An injective random function $f : U \times R \to O_f$ is used to randomize a system-wide unique $u \in U$ with $r \in R$. The output o_f of f is then concatenated to r. Let $U = [0, 2^{l_u} - 1]$, $R = [0, 2^{l_r} - 1]$ and $O_f = [0, 2^{l_{o_f}} - 1]$, where operations are required to be done within $\mathbb{Z}_q$ exclusively, i.e. $U, R, O_f \subseteq [0, q - 1]$. Let $l_q = 1 + l_{o_f} + l_r$, then by Theorem 3.2.1

$$z = f(u, r)2^{l_r} + r$$

is system-wide unique and $z < q$ preserved. Notice that if r and u exist in shared form, then f can be evaluated by using secure multi-party computation. However, one has to be careful which techniques are used, because uniform distribution for z cannot be achieved in $\mathbb{Z}_q$.

Figure 5.9 illustrates the idea in more detail. A narrow box with continuous (or dotted) lines represents a number in reconstructed (or shared) form. The unique input for f consists of a unique identifier UI, a counter c and some random padding ρ_1 and ρ_2. Furthermore, a 0-bit is placed between c and the padding ρ_1, such that c can be incremented at least 2^{l_c} times without destroying UI through an overflow.

During the initialization process (cf. Figure 5.9, outside the main-box), a trusted initializer chooses a unique identifier UI and shares the block $u = UI2^{l_{\rho_2}}$ over a set of players using a (verifiable) secret sharing protocol. The players then jointly generate shares of a secret random counter c (via JVSS) that is restricted to the bit-positions $l_{UI} + l_{\rho_2}$ up to $l_c + l_{UI} + l_{\rho_2} - 1$. This is necessary to ensure that UI is not destroyed when setting $u := u + c$. Establishing c in this form requires the design of a special joint random verifiable secret sharing protocol JVSS, where the secret is restricted to certain bit-positions. The trusted initializer also shares the value $2^{l_{UI} + l_{\rho_2}}$ among the players. Adding this value to u increments the counter c by 1.

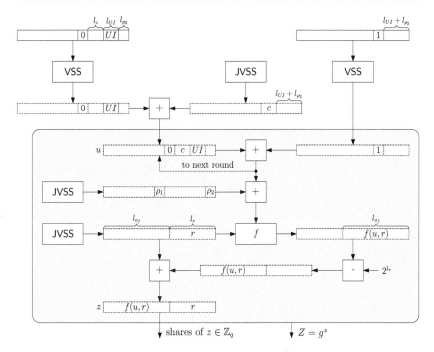

Fig. 5.9: Shared Collision-Free Number Generation.

At the beginning of the key generation protocol (cf. Figure 5.9, inside the main-box), the players jointly add u and $2^{l_{UI}+l_{\rho_2}}$ over the corresponding shares. Moreover, they store the result in u for the next round. Then, they jointly generate the padding ρ, which consists of two blocks ρ_1 and ρ_2, restricted to certain bit-positions, so that computing $\rho + u$ results in $\rho_1||0||c||UI||\rho_2$. This block serves as the unique input for f. It remains to generate shares of a randomizer r, which is restricted to the l_r least significant bits. Now, f can be evaluated using secure multi-party computation, resulting in an output that is restricted to the l_{o_f} least significant bits. Finally, r and o_f need to be concatenated using multi-party techniques. This can be simply done by performing an l_r-bit left-shift (multiplication by the public constant 2^{l_r}) and one shared addition. Since $l_q = 1 + l_{o_f} + l_r$ holds, one gets $z < q$ for the resulting block z, of which each player owns a poly-share. Aside obtaining shares of z, the players have to compute $Z = g^z$. This is not illustrated in Figure 5.9 for simplicity.

Notice that this approach can be realized with security against $t < n/2$ active adversaries if the evaluation of f can be done with the same security level. Keep in mind, however, that u and r are generally not uniformly distributed over an interval so that a suitable probability distribution for the output of f is unlikely to be achieved. In Section 5.5, a very simple instance of f is presented, which provably provides R1 and R3. Unfortunately, the achieved probability distribution of z is not very convincing. Hence, this approach should be considered as an introductory example and motivation for future work, rather than a practical solution.

5.4.3 Shared Collision-Free Pair Generation

If a discrete-logarithm-based cryptosystem is run over $\mathbb{G}_p$, an approach for the shared generation of secrets in $\mathbb{F}_p$ is necessary. With respect to the requirements stated in Section 5.4.1, such secrets need to be system-wide unique. Let $z \in \mathbb{F}_p$ be the secret key and $Z = g^z$ the corresponding public key. To provide uniqueness, it suffices to ensure that the pair $z = (z_1, z_2)$ is completely unique. Let $f : \mathbb{Z}_q \times \mathbb{Z}_q \to \mathbb{Z}_q$ be a function, injective in the first argument for all values of the second argument, $u \in \mathbb{Z}_q$ a system-wide unique number and $r \in_R \mathbb{Z}_q$. Then, from Theorem 3.2.1

$$z = (f(u, r), r)$$

is system-wide unique and $z \in \mathbb{F}_p$. Since f is defined over $\mathbb{Z}_q$, standard multi-party computation can be applied for a shared evaluation among a set of players. Contrary to the previous approach, u and r do not have to lie within a subset of $\mathbb{Z}_q$. This simplifies a shared generation of z. The pair $(z_1, z_2) = (f(u, r), r)$ is never uniformly distributed over $\mathbb{Z}_q^2$ (unlike f is some kind of random oracle), because z_1 and z_2 are mutually dependent. On the contrary, r can be generated using the protocol JPVSS to ensure that its distribution is uniform with respect to $\mathbb{Z}_q$. This has the advantage that given $Z = g^z$, where $z = (f(u, r), r)$, finding z can be shown to be computationally equivalent to solving the Discrete Logarithm Problem. Although secure multi-party computation can be realized in $\mathbb{F}_p$, a protocol for the distributed generation of z and Z has to be realized with respect to multi-party computation in $\mathbb{Z}_q$, because f is defined over $\mathbb{Z}_q$. Nevertheless, the techniques presented in Section 5.3 are still necessary, but for protocols that are run *after* the key generation process.

Figure 5.10 illustrates the idea on an abstract level. Basically, the structure looks quite similar to that of Figure 5.9. The main difference relies on the fact that f can fully operate on elements in $\mathbb{Z}_q$, i.e. is not restricted to subsets of it. The initialization is the same as in the previous section, only the bit-lengths differ. The same holds for the shared generation of the padding, which is chosen such that the block $\rho_1||0||c||UI||\rho_2$ does not exceed $2^{l_q - 1}$. Another difference is that the random part r can be generated by a standard joint random verifiable secret sharing protocol JVSS, such as JPVSS. As a consequence, r is provably uniformly distributed in $\mathbb{Z}_q$. The function f can be

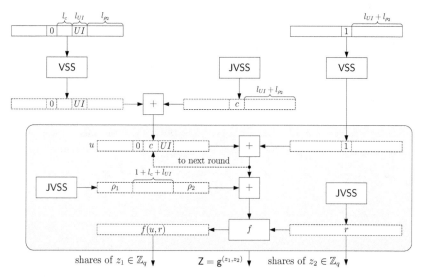

Fig. 5.10: Shared Collision-Free Pair Generation.

evaluated using secure multi-party computation in $\mathbb{Z}_q$. The output of the protocol are poly-shares of z_1 and poly-shares of z_2. As shown in Section 5.3, such poly-shares can be interpreted as a fusion-share in $\mathbb{F}_p$. Aside computing shares, the public key $\mathsf{Z} = \mathsf{g}^z$ needs to be computed which is not illustrated in Figure 5.10 for simplicity.

According to R2, one must not learn anything about z, except for what is implied by Z. Since $z = (f(u, r), r)$ we have a dependency between the two components of z. So, the security at least relies on the uncertainty about r. Section 5.6.4 shows that computing r from Z is computationally equivalent to solving the Discrete Logarithm Problem. A consequence of this result is that secret keys of the form $(f(u, r), r)$ are secure even if the output of f follows a bad probability distribution.

5.4.4 Shared Multiplication of Unique Primes

This section sketches an exotic approach for the generation of secret keys in $\mathbb{Z}_q$. Let $\{p_i\}_{i \in I(\mathcal{P})}$ be a set of primes, where each p_i is the output of a collision-free number generator. Moreover, let $\sum_{i=1}^{n} l_{p_i} < l_q$. Then, from Theorems 3.2.1 and 5.2.1

$$z = \prod_{i \in I(\mathcal{Q})} p_i$$

is system-wide unique and $z < q$ preserved. Notice that there is no need to actually *share* a collision-free number generator. Instead, each player is provided with an ordinary one used locally. Secret key generation in $\mathbb{Z}_q$ requires that $p_1 \cdot \ldots \cdot p_n <$ q. Hence, each prime p_i must be significantly smaller than q, otherwise a reduction modulo q takes place and the product z is no longer unique. As far as security is concerned, the bit-length of the product of any $t + 1$ of the n primes must be large enough with respect to a certain probability distribution to keep computing discrete logarithms hard. Notice that z is a product that is *not* reduced modulo q and hence follows a probability distribution that approaches *log-normal distribution* [LSA01] over a sub-interval of $[0, q - 1]$. Thus, when analyzing the security, one has to focus on the uncommon variants of the Discrete Logarithm Problem defined in Section 5.2.1.

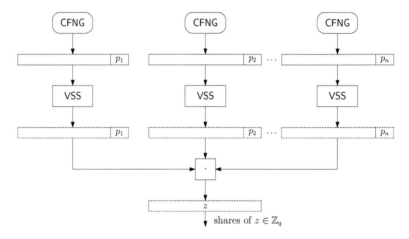

Fig. 5.11: Shared Multiplication of Unique Primes.

Figure 5.11 illustrates the generation process. The players are provided with a collision-free number generator CFNG. At the beginning, each player P_i generates a system-wide unique prime p_i and shares it over the other players using an appropriate verifiable secret sharing protocol. Hence, each player holds a share of each other prime. Then, the players perform several runs of a shared multiplication protocol. To keep the number of sent message minimal, independent multiplications are done in parallel. So, the number of sequential runs of the multiplication protocol is approximately $log_2(n)$. Since each instance uses the collision-free number generator locally, the protocol is at most secure against passive adversaries if no trusted environment is used. A secret sharing scheme can be given where each player is convinced that a shared secret is a *prime* that lies in a certain interval [LNV99]. However, it is challenging to prove that a shared secret is a *system-wide unique prime*. Two possible alternatives are:

1. All steps are done in software and hence security holds against passive adversaries.

2. The steps of CFNG are done in a trusted environment. All other steps are done in software for which the security holds against active adversaries.

Security against active adversaries is more relevant in practice. Thus, the practical part of this chapter focuses on a solution for the second case (cf. Section 5.7). Requiring a trusted environment is not impractical. Since every player needs to have access to a collision-free number generator, it is self-evident to use smartcards if the protocol is not run often. Otherwise, one has to use an appropriate crypto-hardware.

5.4.5   Common Assumptions

For all described variants it is assumed that an appropriate security infrastructure exists to provide confidential, integrity-protected and authentic communication between the players. Furthermore, the players must be connected to a secure broadcast channel, i.e. if a player sends a message using this channel, all other players get exactly this message immediately. For dishonest players we expect that they behave staticly, which means that they decide at the beginning of the protocol if they are honest or dishonest. Computational assumptions for the adversary model are given for each protocol separately in the corresponding section.

5.5   Distributed Key Generation Protocol DKG1

In this section, a practical solution for shared collision-free number generation, as described in Section 5.4.2, is given. Therefore, the function f is defined as follows:

$$f : \{0,1\}^{l_u} \times \{0,1\}^{l_r} \to \{0,1\}^{l_{o_f}}, \quad f(u,r) = ur \text{ MOD } q$$

where $l_q = 1 + l_{o_f} + l_r$ holds. The computation MOD q has no real effect, but denotes that the function can be evaluated in $\mathbb{Z}_q$. For the shared generation of c, r and the random padding ρ, a special protocol is necessary, which is described in the following.

5.5.1   Shared Generation of Normally Distributed Blocks

JPVSS can be used for the shared generation of a random secret which is *uniformly* distributed over $\mathbb{Z}_q$. For the approach presented in the current section, a special variant of JPVSS is necessary where the players jointly generate an element $x \in \mathbb{Z}_q$, such that the bits of one or more certain blocks are chosen at random, and all other bits are verifiably set to 0. For simplicity, the case where the players want to generate poly-shares of $x \in [0, 2^{l_x} - 1]$ in $\mathbb{Z}_q$ is considered, where $l_x < l_q$. Thus, x is defined as the sum of at least $t + 1$ and at most n random numbers submitted by the players. The length

---

Protocol JPVSS2

Input: BL

Output (P_i): $([x_i]^q, [x_i']^q), \{[x_{ji}]^q, [x_{ji}']^q, \widetilde{X}_j\}_{j \in I(\mathcal{Q})}, \{r_{ik}\}_{k \in \mathcal{T}}$

1. Each P_i chooses $\{x_{ij} \in_R \{0,1\}^{len_j - \lceil log_2(n)\rceil}\}_{j=1}^{|BL|}$ and computes $x_i = \sum_{j=1}^{|BL|} x_{ij} 2^{pos_j}$.

2. Then, each P_i shares x_i over the other players using PVSS2, where $B = \{j \mid pos \leq j < pos + len - \lceil log_2(n)\rceil \wedge (pos, len) \in BL\}$. Hence, P_j gets $\{([x_{ij}]^q, [x_{ij}']^q)\}_{i \in I(\mathcal{Q})}$ and $\{\widetilde{X}_{ik}\}_{i,k \in I(\mathcal{Q})}$ as output. Moreover, he stores the set of his own sharing coefficients $\{r_{jk}\}_{k \in \mathcal{T}}$.

3. Now each player P_j computes $[x_j]^q = \sum_{i \in I(\mathcal{Q})}[x_{ij}]^q$ and $[x_j']^q = \sum_{i \in I(\mathcal{Q})}[x_{ij}']^q$.

4. Finally, each player computes the set $\{\widetilde{X}_j = \prod_{i \in I(\mathcal{Q})}\widetilde{X}_{ij}\}_{j \in I(\mathcal{Q})}$.

Fig. 5.12: Joint Pedersen Verifiable Sharing of Normally Distributed Secrets.

of the input x_i of player P_i must not exceed $l_x - \lceil log_2(n)\rceil$, otherwise $\sum_{i \in I(\mathcal{Q})} x_i \geq 2^{l_x}$ might occur. Contrary to as done in JPVSS, $\sum_{i \in I(\mathcal{Q})} x_i$ is *not* reduced modulo a number. Hence, x is *not* uniformly distributed over a certain interval. However, the larger n is, the more x approaches a *normal distribution* [LSA01] over $[0, 2^{l_x} - 1]$ if each x_i is chosen uniformly over $[0, 2^{l_x - \lceil log_2(x)\rceil} - 1]$. Notice that t of n dishonest players may not follow the protocol specification and publish their secret inputs. Then, the uncertainty about x lies in the $l_x - \lceil log_2(t)\rceil$ bits that are normally distributed. If x has been generated by a protocol which provides unconditional security, then the best strategy of an attacker is to guess the remaining $l_x - \lceil log_2(t)\rceil$ bits with respect to the knowledge that the corresponding block is normally distributed. In our case, however, the key generation protocol contains steps for the computation of a public key, that is a D-commitment of the corresponding secret key.

According to [BT00, Tes01], the best known attack against the Discrete Logarithm Problem with respect to a normally distributed secret x, is the baby-giant-step algorithm. The expected running time is $O(\sqrt{E(X)})$, where $E(X) = \sum_{i \in I(\mathcal{Q})} E(X_i)$ is the expected value for the sum of random variables and $E(X_i)$ the expected value for the i-th random variable. In this case the Probability Distribution Discrete Logarithm Problem becomes the Expected Value Discrete Logarithm Problem (cf. Section 5.2.1). To keep the running time of the baby-giant-step algorithm impractically long, one has to choose l_x sufficiently large. It seems as if Pollard's rho algorithm cannot be used to take advantage of the known non-uniform distribution of a discrete logarithm [Tes01].

Figure 5.12 outlines protocol JPVSS2. The public input is BL, which is a set containing pairs of the form (pos, len), specifying that the bits from bit-position pos up to bit-position $pos + len - 1$ are chosen at random and all other bits are equal to zero. At the beginning of JPVSS2, each player P_i chooses a number $x_{ij} \in \{0,1\}^{len_j - \lceil log_2(n) \rceil}$ at random for each pair $(pos_j, len_j) \in BL$. Afterwards, he computes $x_i = \sum_{j=1}^{|BL|} x_{ij} 2^{pos_j}$, i.e. concatenates all blocks as specified by BL. Next, P_i shares x_i over the other players using PVSS2, where B is chosen consistent with BL, i.e. B contains the indices of all bits that are chosen at random. The remaining steps are analogous to protocol JPVSS. Notice that uniformly distributed shares and P-commitments are sent over the network. Up to t shares and all P-commitments, actually reveal no information about x. Indeed up to t players may reveal their secret inputs. Thus, as mentioned, the uncertainty relies on x being normally distributed over $[0, 2^{l_x - \lceil log_2(t) \rceil} - 1]$.

5.5.2   Shared Collision-Free Number Generation

To ensure uniqueness $l_q = 1 + l_{o_f} + l_r$, $l_{o_f} = l_u + l_r$ and $l_u = l_{\rho_1} + 1 + l_c + l_{UI} + l_{\rho_2}$ is required. At this point no suggestions concerning the practical choice of the bit-lengths are made. Protocol DKG1 requires a special initialization protocol, which is given in Figure 5.13. An initializer chooses a unique identifier UI of length l_{UI} and performs an l_{ρ_2}-bit left-shift. The resulting value u and $2^{l_{UI}+l_{\rho_2}}$ are both shared among the players using PVSS. The use of verifiable secret sharing is not necessary here to check if the initializer behaves honestly, but to provide each player with authentic P-commitments and shares for later use. The i-th share of $2^{l_{UI}+l_{\rho_2}}$ is denoted as $[1_i]^q$ since it is later used to increment the counter c by 1. In step 3, the players jointly generate shares of a random counter c which is restricted to the bit-positions $l_{UI} + l_{\rho_2}$ up to $l_c + l_{UI} + l_{\rho_2} - 1$. Therefore they must run JPVSS2, where BL is defined appropriately. Then, the players run a multi-party addition to compute $u := u + c$ using the corresponding shares. Notice that none of the bits gets lost.

When participating in DKG1, each P_i inputs the values he received during initialization. Then, in step 1, the players increment the counter in u by 1. In step 2, they jointly generate the two secrets ρ and r using two runs of protocol PVSS2. Since those runs are mutually independent, they can be done in parallel. The secret ρ contains random padding at the beginning and at the end of the unique block u, such that $(u + \rho) = \rho_1 ||0||c||UI||\rho_2$ and $l_{(u+\rho)} \leq l_u$ holds. Hence, BL needs to be formed properly as stated in step 2a. The randomizer r needs to be restricted to the l_r least significant bits (cf. step 2b). The inputs for f are now the unique part $(u + \rho)$ and the random part r. Since $f(u,r) = ur$, the two secrets u and r are just multiplied via VMUL resulting in shares of o_f. For the multiplication it suffices to run VMUL without any modifications because it has already been ensured that the bits are set properly during the generation of u, ρ and r. Finally, it remains to compute $o_f || r$. Therefore, o_f needs to be shifted to the left by l_r positions. This can be done through

---

Protocol InitDKG1

Output (P_i): $([u_i]^q, [u_i']^q), ([1_i]^q, [1_i']^q), \{\widetilde{U}_j, \widetilde{1}_j\}_{j \in I(\mathcal{Q})}$

1. The initializer chooses $UI \in \{0,1\}^{l_{UI}}$ and computes $u = UI2^{l_{\rho_2}}$.

2. Then, u and $2^{l_{UI}+l_{\rho_2}}$ are shared among the players using PVSS, i.e. each P_i receives $([u_i]^q, [u_i']^q), ([1_i]^q, [1_i']^q)$ and $\{\widetilde{U}_j, \widetilde{1}_j\}_{j \in I(\mathcal{P})}$ as output.

3. The players run JPVSS2 to compute shares of a counter c, where $BL = \{(l_{UI} + l_{\rho_2}, l_c)\}$. The (used) output of P_i is $([c_i]^q, [c_i']^q)$ and $\{\widetilde{C}_j\}_{j \in I(\mathcal{Q})}$.

4. Each P_i sets $[u_i]^q = [u_i]^q + [c_i]^q$, $[u_i']^q = [u_i']^q + [c_i']^q$ and $\{\widetilde{U}_j = \widetilde{U}_j \widetilde{C}_j\}_{j \in I(\mathcal{Q})}$.

Protocol DKG1

Input (P_i): $([u_i]^q, [u_i']^q), ([1_i]^q, [1_i']^q), \{\widetilde{U}_j, \widetilde{1}_j\}_{j \in I(\mathcal{Q})}$

Output (P_i): $[z_i]^q, ([u_i]^q, [u_i']^q), \{Z_j, \widetilde{U}_j\}_{j \in I(\mathcal{Q})}, Z$

1. Each P_i computes $[u_i]^q = [u_i]^q + [1_i]^q$, $[u_i']^q = [u_i']^q + [1_i']^q$ and $\{\widetilde{U}_j = \widetilde{U}_j \widetilde{1}_j\}_{j \in I(\mathcal{Q})}$.

2. The players then run the following two steps in parallel:

   a) Shares of a secret padding ρ are generated via JPVSS2, where $BL = \{(1 + l_c + l_{UI} + l_{\rho_2}, l_{\rho_1}), (0, l_{\rho_2})\}$. The output of player P_i is $([\rho_i]^q, [\rho_i']^q)$ and $\{\widetilde{\rho}_j\}_{j \in I(\mathcal{Q})}$.

   b) Shares of a secret r are generated via JPVSS2, where $BL = \{(0, l_r)\}$. The output of player P_i is $([r_i]^q, [r_i']^q)$ and $\{\widetilde{R}_j\}_{j \in I(\mathcal{Q})}$.

3. The players perform a shared multiplication over the shares of $(u + \rho)$ and r. Hence, each P_i inputs $([u_i]^q + [\rho_i]^q, [u_i']^q + [\rho_i']^q), ([r_i]^q, [r_i']^q)$ and $\{\widetilde{U}_j \widetilde{\rho}_j, \widetilde{R}_j\}_{j \in I(\mathcal{Q})}$ to VMUL and gets $([o_{f_i}]^q, [o_{f_i}']^q)$ and $\{\widetilde{O}_{f_j}\}_{j \in I(\mathcal{Q})}$ as output.

4. Each P_i computes $[z_i]^q = [o_{f_i}]^q 2^{l_r} + [r_i]^q$, $[z_i']^q = [o_{f_i}']^q 2^{l_r} + [r_i']^q$ and the set of the corresponding commitments $\{\widetilde{Z}_j = \widetilde{O}_{f_j}^{2^{l_r}} \widetilde{R}_j\}_{j \in I(\mathcal{Q})}$.

5. Each P_i broadcasts $Z_i = g^{[z_i]^q}$ and a Σ-proof τ_i, that $dlog_g(Z_i) = dlog_{g_1}(\widetilde{Z}_i g_2^{-[z_i']^q})$.

6. For each j, every player P_i checks if $V(\tau_j, (Z_j, \widetilde{Z}_j)) = 1$ and broadcasts a complaint against P_j if not. If there are more than t complaints, then P_j is disqualified.

7. Finally, each player computes the public key $Z = \prod_{i \in I(\mathcal{Q})} Z_i^{\lambda_i}$.

Fig. 5.13: Shared Collision-Free Number Generation.

a local multiplication by the public constant 2^{l_r}, i.e. no communication is necessary. The resulting block is added to r. The corresponding commitments can be computed straightforwardly due to the homomorphic property. Finally, the public key $Z = g^z$ needs to be computed. Therefore, each player broadcasts a D-commitment of his share and proves in zero-knowledge, that it corresponds to his P-commitment. This can be done using a standard non-interactive Σ-proof for the equality of discrete logarithms:

each P_i computes $c_i = \mathcal{H}(g^{\alpha_{i1}}||g_1^{\alpha_{i1}} g_2^{\alpha_{i2}})$, where $\alpha_{i1}, \alpha_{i2} \in_R \mathbb{Z}_q$. Moreover, he computes $s_{i1} = \alpha_{i1} - c_i[z_i]^q$ and $s_{i2} = \alpha_{i2} - c_i[z_i']^q$. Then he sets $\tau_i := (c_i, s_{i1}, s_{i2})$. Given τ_i and the pair $(Z_i, \widetilde{Z}_i)$, any player can verify τ_i with $V : T \times \mathbb{G}_q^2 \to \{0, 1\}$, defined as follows:

$$V(\tau_i, (Z_i, \widetilde{Z}_i)) := \begin{cases} 0 & \text{if } \tau_{i1} \neq \mathcal{H}(Z_i^{\tau_{i1}} g^{\tau_{i2}}||\widetilde{Z}_i^{\tau_{i1}} g_1^{\tau_{i2}} g_2^{\tau_{i3}}) \\ 1 & \text{otherwise} \end{cases}$$

It can be shown that the above proof is a non-interactive Σ-proof. This can be done analogously to Section 5.2.3.2. The D-commitments, for which the proofs have been accepted, are then used to generate the public key over a hidden interpolation.

5.5.3 Correctness

To show the correctness of protocol DKG1, one has to prove that (a), (b) and (c) of R1 hold. All parties listen to a broadcast channel and hence every player determines the same set $\mathcal{Q}$. It is assumed that during the init-protocol the initializer behaved honestly, such that each player is provided with correct shares and commitments.

Theorem 5.5.1 *With the assumption that solving the Discrete Logarithm Problem is hard in $\mathbb{G}_q$, protocol DKG1 fulfills R1 if at most $t < n/2$ players are actively dishonest.*

Proof. It has to be shown that (a), (b) and (c) of R1 hold.

(a) In step 1, each player locally computes $[u_i]^q$ and P-commitments of all other shares. Since no communication takes place, the values are correct for at least $t+1$ honest players. In step 2, the players jointly generate the padding ρ for u and the random part r of the secret key. Since protocol PVSS2 is used, it is guaranteed that each honest player owns a share of ρ, a share of r and the P-commitments of all shares in the system. Moreover, PVSS2 ensures that ρ and r are formed correctly. Notice that actively dishonest parties are disqualified and the shares of the honest players uniquely define ρ and r. In step 3, $(\rho + u)$ and r are multiplied over the corresponding shares. Using protocol VMUL, actively dishonest parties can be identified and disqualified ensuring that only honest players hold valid shares. Per assumption, $t < n/2$ holds and so up to t players can deviate from the specified protocol steps, however, for $t+1$ shares correctness is in any case guaranteed.

(b) It suffices to show that correct shares can be distinguished from incorrect ones: a share $[z_i]^q$ is correct, if Z_i and $\widetilde{Z}_i$ are both commitments of the same secret $[z_i]^q$. This holds with computational uniqueness, since P-commitments are computational binding. Per assumption, $dlog_{g_1}(g_2)$ is unknown and so (b) holds.

(c) Malicious parties can be uniquely identified and disqualified. Thus, their inputs and corresponding commitments can be identified and invalidated locally by every honest player. Since the public key Z is uniquely computed using the commitments of the honest players, every of these players holds the same public key.   □

5.5.4   Secrecy

It is obvious that, compared to the typical setting where keys are uniformly distributed, secret keys generated through DKG1 provide a dissatisfying distribution. As a consequence, a security analysis is sketched in the following instead of giving formal proofs.

The poly-shares of u and r are hidden using P-commitments which are perfect hiding. Hence, one cannot learn anything about u or r from the commitments. An adversary having access to t shares of u or r cannot learn anything new, since poly-sharing is used. If t random inputs for the generation of r are publicly known, then the uncertainty about r relies on the sum of the remaining $t + 1$ inputs of the other honest players. The same argument holds the generation of ρ, since it is computed analogously. The uncertainty about z relies on at least three blocks of length $l_r - \lceil log_2(t) \rceil$, $l_{\rho_1} - \lceil log_2(t) \rceil$ and $l_{\rho_2} - \lceil log_2(t) \rceil$, that are normally distributed respectively. We recommend to choose l_r, l_{ρ_1} and l_{ρ_2} sufficiently large to render the Probability Distribution Discrete Logarithm Problem hard for a poly-bounded adversary. The involved non-interactive Σ-proof obviously reveals no information about the shares.

5.5.5   Uniqueness

It can be proven that the requirement for uniqueness is fulfilled for at least 2^{l_c} runs:

Theorem 5.5.2 *With the assumption that solving the Discrete Logarithm Problem is hard in $\mathbb{G}_q$, protocol DKG1 fulfills R3 if at most $t < n/2$ players are actively dishonest.*

Proof. To show that uniqueness is fulfilled, one has to prove that $(\rho + u)r2^{l_r} + r$ is system-wide unique. Since multiplication in $\mathbb{Z}_q$ is bijective, $(\rho + u)r2^{l_r} + r$ is unique if $(\rho + u)$ is unique (cf. Theorem 3.2.1) and r restricted to the l_r least significant bits. The latter holds due to the run of JPVSS2. It remains to show that $(\rho + u)$ is unique. Notice that the initialization protocol guarantees that $[u_i]^q$ is a poly-share of $u = (c||UI)2^{l_{\rho_2}}$, where UI is a system-wide unique identifier. In step 1, the players jointly increment u using shared addition of u and $2^{l_{UI}+l_{\rho_2}}$. Since no communication takes place, u remains unique. In step 2a the players jointly generate shares of ρ. Protocol JPVSS2 ensures that the random bits in ρ are restricted to the bit-positions 0 up to $l_{\rho_2} - 1$ and the bit-positions $1 + l_c + l_{UI} + l_{\rho_2}$ up to $l_{\rho_1} + l_c + l_{UI} + l_{\rho_2}$. Hence, computing $(\rho + u)$ results in $\rho_1||0||c||UI||\rho_2$ which still preserves uniqueness. Since c is initialized randomly, the key generation protocol can be repeated 2^{l_c} times without having an overflow that destroys the uniqueness.   □

5.6 Distributed Key Generation Protocol DKG2

In this section, a practical solution for shared collision-free pair generation is given (cf. Section 5.4.3). The function f is defined as follows:

$$f : \mathbb{Z}_q \times \mathbb{Z}_q \to \mathbb{Z}_q, \quad f(u, r) = ur$$

Regarding the size of u and r, both must be elements of $\mathbb{Z}_q$. Although this approach looks similar to the previous one, there is an intrinsic difference: the players can use JPVSS to generate shares of the random part r, which is *uniformly distributed* in $\mathbb{Z}_q$. The system-wide unique element u is generated analogously to the previous situation. Only the bit-lengths differ, because one can use a longer padding.

Since f is based on $\mathbb{Z}_q$ exclusively, standard protocols for secure multi-party computation can be used to evaluate f. The goal of the key generation protocol is to provide each player P_i with a fusion-share $\langle z_i \rangle^q$ of $z = (f(u, r), r)$ and the public key $Z = g^z$, which requires some additional steps. For efficiency reasons, a special shared multiplication protocol is given and described in the following.

5.6.1 Special Verifiable Shared Multiplication

Let a and b be two secrets that have been shared among n players by using PVSS. Hence, each player P_i is provided with the four shares $[a_i]^q$, $[a_i']^q$, $[b_i]^q$ and $[b_i']^q$, and the P-commitments $\widetilde{A}_i = g_1^{[a_i]^q} g_2^{[a_i']^q}$ and $\widetilde{B}_i = g_1^{[b_i]^q} g_2^{[b_i']^q}$. The players want to compute shares of the secret pair (ab, b) without intermediate reconstruction of a, b or ab. The goal is that after completing the protocol each player P_i posses a fusion-share $([c_i]^q, [b_i]^q)$, where $[c_i]^q$ is a poly-share of ab and $[b_i]^q$ a novel poly-share of b. The term "novel" is used as the suggested protocol requires a re-sharing of b with respect to the technique described in Section 5.2.4.3. Thereby, b itself is not changed.

At the beginning of VMUL2, each player locally multiplies $[a_i]^q$ by $[b_i]^q$ resulting in the share $[d_i]^q$, that lies on a $2t$-degree polynomial. Similarly to VMUL, it is necessary to re-share $[d_i]^q$ to get a t-degree random share $[c_i]^q$ of ab. In VMUL, the additional goal is that each player is finally provided with the P-commitments of each share of ab. In the current situation, each player needs to get an FD-commitment of each pair $([c_i]^q, [b_i]^q)$ instead. This can be efficiently achieved if $[b_i]^q$ is also re-shared using a t-degree polynomial. Thus, in step 2a, each P_i chooses random coefficients $r_{d_{i1}}, r_{b_{i1}}, \ldots, r_{d_{it}}, r_{b_{it}} \in \mathbb{Z}_q$ and sets $r_{d_{i0}} := [d_i]^q$ and $r_{b_{i0}} := [b_i]^q$. The player then forms the sharing polynomials $g_{[d_i]^q}$ and $g_{[b_i]^q}$. In step 2c, P_i broadcasts the FD-commitments $\{R_{ik}\}_{k \in \mathcal{T}}$, where $R_{ik} = g^{(r_{d_{ik}}, r_{b_{ik}})}$. Hence, each receiver can compute the FD-commitment D_{ij} of each fusion-share $([d_{ij}]^q, [b_{ij}]^q)$ of the secret $([d_i]^q, [b_i]^q)$ using a hidden evaluation of the

---

Protocol VMUL2

Input (P_i): $([a_i]^q, [a_i']^q)$, $([b_i]^q, [b_i']^q)$, $\{\widetilde{A}_j, \widetilde{B}_j\}_{j \in I(\mathcal{Q})}$

Output (P_i): $([c_i]^q, [b_i]^q)$, $\{\mathsf{C}_j\}_{j \in I(\mathcal{Q})}$

1. Each P_i computes $[d_i]^q = [a_i]^q [b_i]^q$.

2. Then each P_i runs the following re-sharing procedure:

   a) P_i chooses the coefficients $r_{d_{i1}}, r_{b_{i1}}, \ldots, r_{d_{it}}, r_{b_{it}} \in_R \mathbb{Z}_q$ and sets $r_{d_{i0}} := [d_i]^q$
      and $r_{b_{i0}} := [b_i]^q$. Then he forms the sharing-polynomials $g_{[d_i]^q}(x)$ and $g_{[b_i]^q}(x)$.

   b) Next, P_i forms $\{\mathsf{R}_{ik} = \mathsf{g}^{(r_{d_{ik}}, r_{b_{ik}})}\}_{k \in \mathcal{T}}$ and generates a non-interactive proof τ_i
      to show that R_{i0} is an FD-commitment of $([a_i]^q [b_i]^q, [b_i]^q)$.

   c) P_i broadcasts $\{\tau_i, \mathsf{R}_{ik}\}_{k \in \mathcal{T}}$ and sends $([d_{ij}]^q, [b_{ij}]^q) = (g_{[d_i]^q}(j), g_{[b_i]^q}(j))$ to P_j.

   d) Each player computes $\{\mathsf{D}_{ij} = \prod_{k \in \mathcal{T}} \mathsf{R}_{ik}^{(j^k, 0)}\}_{i,j \in I(\mathcal{P})}$.

   e) Then each P_j does the following verification:

   $$\mathsf{D}_{ij} \overset{?}{=} \mathsf{g}^{([d_{ij}]^q, [b_{ij}]^q)}, \quad V(\tau_i, (\widetilde{A}_i, \widetilde{B}_i, \mathsf{R}_{i0})) \overset{?}{=} 1$$

   If any of the two verifications fails, he broadcasts a complaint against P_i. If
   there are more than t complaints, P_i is disqualified. If the are less than t
   complaints with respect to the left verification, P_i broadcasts $([d_{ij}]^q, [b_{ij}]^q)$ for
   each complainer P_j. If the pair is correct P_j is disqualified, otherwise P_i.

3. Finally, each P_j computes $[c_j]^q = \sum_{i \in I(\mathcal{Q})} [d_{ij}]^q \lambda_i$, $[b_j]^q := \sum_{i \in I(\mathcal{Q})} [b_{ij}]^q \lambda_i$ and the
   set of FD-commitments of all other shares: $\{\mathsf{C}_k = \prod_{i \in I(\mathcal{Q})} \mathsf{D}_{ik}^{(\lambda_i, 0)}\}_{k \in I(\mathcal{Q})}$.

Fig. 5.14: Special Verifiable Shared Multiplication.

sharing-polynomials $g_{[d_i]^q}(j)$ and $g_{[b_i]^q}(j)$ as follows:

$$\mathsf{D}_{ij} = \prod_{k \in \mathcal{T}} \mathsf{R}_{ik}^{(j^k, 0)}$$

It can be shown that $\mathsf{D}_{ij} = \mathsf{g}^{([d_{ij}]^q, [b_{ij}]^q)}$ must hold, if each R_{ik} is formed correctly:

$$\mathsf{D}_{ij} = \prod_{k \in \mathcal{T}} \mathsf{R}_{ik}^{(j^k, 0)} = (\prod_{k \in \mathcal{T}} (g_1^{r_{d_{ik}}} g_2^{-r_{b_{ik}}})^{j^k}, \prod_{k \in \mathcal{T}} (g_1^{r_{b_{ik}}} g_2^{r_{d_{ik}}})^{j^k})$$

$$= (g_1^{\sum_{k \in \mathcal{T}} r_{d_{ik}} j^k} g_2^{-\sum_{k \in \mathcal{T}} r_{b_{ik}} j^k}, g_1^{\sum_{k \in \mathcal{T}} r_{b_{ik}} j^k} g_2^{\sum_{k \in \mathcal{T}} r_{d_{ik}} j^k})$$

$$= (g_1^{[d_{ij}]^q} g_2^{-[b_{ij}]^q}, g_1^{[b_{ij}]^q} g_2^{[d_{ij}]^q}) = \mathsf{g}^{([d_{ij}]^q, [b_{ij}]^q)}$$

In the next step, each P_j has to check that each received pair $([d_{ij}]^q, [b_{ij}]^q)$ is consistent
with the corresponding FD-commitment D_{ij}, as computed in step 2d. If this is not
the case, then P_j has to broadcast a complaint against P_i.

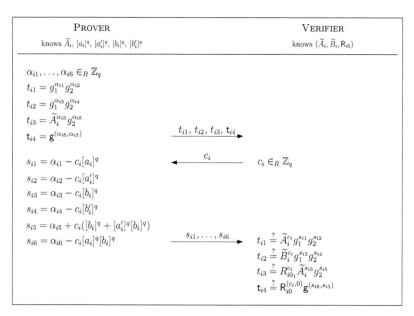

Fig. 5.15: Proof that R_{i0} is the FD-commitment of $([a_i]^q[b_i]^q, [b_i]^q)$.

So far, every P_j is convinced that he owns a fusion-share of the *same secret* as each other player. However, one cannot be sure that each shared secret is formed correctly. Hence, when broadcasting the FD-commitments $R_{i0}, \ldots, R_{it}$, P_i also has to prove that R_{i0} is indeed an FD-commitment of the secret pair $([a_i]^q[b_i]^q, [b_i]^q)$. This can be done using a similar Σ-proof to the one given in Section 5.2.3.2, since every receiver has authentic access to $\widetilde{A}_i = g_1^{[a_i]^q} g_2^{[a_i']^q}$ and $\widetilde{B}_i = g_1^{[b_i]^q} g_2^{[b_i']^q}$. The interactive proof can be found in Figure 5.15. Let $R_{i0} = (R_{i0_1}, R_{i0_2})$. Completeness for the first two verifications is trivial. For the third and forth verification, completeness can be verified easily:

$$R_{i0_1}^{c_i} \widetilde{A}_i^{s_{i3}} g_2^{s_{i5}} = (g_1^{[a_i]^q[b_i]^q} g_2^{-[b_i]^q})^{c_i} (g_1^{[a_i]^q} g_2^{[a_i']^q})^{\alpha_{i3} - c_i[b_i]^q} g_2^{\alpha_{i5} + c_i([b_i]^q + [a_i']^q[b_i]^q)}$$

$$= g_1^{[a_i]^q[b_i]^q c_i} g_2^{-[b_i]^q c_i} g_1^{[a_i]^q(\alpha_{i3} - c_i[b_i]^q)} g_2^{[a_i']^q(\alpha_{i3} - c_i[b_i]^q)} g_2^{\alpha_{i5} + c_i[b_i]^q + c_i[a_i']^q[b_i]^q}$$

$$= g_1^{[a_i]^q \alpha_{i3}} g_2^{[a_i']^q \alpha_{i3}} g_2^{\alpha_{i5}} = (g_1^{[a_i]^q} g_2^{[a_i']^q})^{\alpha_{i3}} g_2^{\alpha_{i5}} = \widetilde{A}_i^{\alpha_{i3}} g_2^{\alpha_{i5}} = t_{i3}$$

$$R_{i0}^{(c_i, 0)} \mathbf{g}^{(s_{i6}, s_{i3})} = (\mathbf{g}^{([a_i]^q[b_i]^q, [b_i]^q)})^{(c_i, 0)} \mathbf{g}^{(\alpha_{i6} - c_i[a_i]^q[b_i]^q, \alpha_{i3} - c_i[b_i]^q)}$$

$$= \mathbf{g}^{([a_i]^q[b_i]^q, [b_i]^q)(c_i, 0)} \mathbf{g}^{(\alpha_{i6}, \alpha_{i3}) - (c_i[a_i]^q[b_i]^q, c_i[b_i]^q)}$$

$$= \mathbf{g}^{([a_i]^q[b_i]^q, [b_i]^q)(c_i, 0)} \mathbf{g}^{(\alpha_{i6}, \alpha_{i3}) - (c_i, 0)([a_i]^q[b_i]^q, [b_i]^q)} = \mathbf{g}^{(\alpha_{i6}, \alpha_{i3})} = t_{i4}$$

Special soundness and honest-verifier zero-knowledge can be shown straightforwardly.

In step 2b of VMUL2, each P_i broadcasts a non-interactive version τ_i of the protocol stated in Figure 5.15, which can be generated as follows: P_i computes $t_{i1}, t_{i2}, t_{i3}, t_{i4}$ and $s_{i1}, \ldots, s_{i6}$ as done in Figure 5.15. Thereby, he sets $c_i := \mathcal{H}(t_{i1}||t_{i2}||t_{i3}||t_{i4})$ and $\tau_i := (c_i, s_{i1}, \ldots, s_{i6})$. Given τ_i and the triple $(\widetilde{A}_i, \widetilde{B}_i, \mathsf{R}_{i0})$, one can verify the correctness of τ_i with the function $V : T \times (\mathbb{G}_q \times \mathbb{G}_q \times \mathbb{G}_p) \to \{0, 1\}$, which is defined as follows:

$$
V(\tau_i, (\widetilde{A}_i, \widetilde{B}_i, \mathsf{R}_{i0})) := \begin{cases} 0 & \text{if } \tau_{i1} \neq \mathcal{H}(\widetilde{A}_i^{\tau_{i1}} g_1^{\tau_{i2}} g_2^{\tau_{i3}} || \widetilde{B}_i^{\tau_{i1}} g_1^{\tau_{i4}} g_2^{\tau_{i5}} || \\ & \quad R_{i0_1}^{\tau_{i1}} \widetilde{A}_i^{\tau_{i4}} g_2^{\tau_{i6}} || \mathsf{R}_{i0}^{(\tau_{i1},0)} \mathbf{g}^{(\tau_{i7}, \tau_{i4})}) \\ 1 & \text{otherwise} \end{cases}
$$

If the verification fails, the verifier sends a complaint against P_i. If there are more than t complaints, then P_i is disqualified. Notice that the complaints concerning the left verification and the right verification of step 2e can be carried out in parallel.

In step 3 of VMUL2, each player P_i locally computes his fusion-share of (ab, b) via component-wise Lagrange interpolation. Moreover, he computes the FD-commitments of all other fusion-shares straightforwardly with a hidden interpolation:

$$
\begin{aligned}
\mathsf{C}_k &= \prod_{i \in I(\mathcal{Q})} \mathsf{D}_{ik}^{(\lambda_i, 0)} = \left(\prod_{i \in I(\mathcal{Q})} (g_1^{[d_{ik}]^q} g_2^{-[b_{ik}]^q})^{\lambda_i}, \prod_{i \in I(\mathcal{Q})} (g_1^{[b_{ik}]^q} g_2^{[d_{ik}]^q})^{\lambda_i} \right) \\
&= \left(g_1^{\sum_{i \in I(\mathcal{Q})} [d_{ik}]^q \lambda_i} g_2^{-\sum_{i \in I(\mathcal{Q})} [b_{ik}]^q \lambda_i}, g_1^{\sum_{i \in I(\mathcal{Q})} [b_{ik}]^q \lambda_i} g_2^{\sum_{i \in I(\mathcal{Q})} [d_{ik}]^q \lambda_i} \right) \\
&= \left(g_1^{[c_k]^q} g_2^{-[b_k]^q}, g_1^{[b_k]^q} g_2^{[c_k]^q} \right) = \mathbf{g}^{([c_k]^q, [b_k]^q)}
\end{aligned}
$$

5.6.2 Shared Collision-Free Pair Generation

A protocol must be designed which provides each player with a fusion-share of a unique secret z and the corresponding FD-commitment $\mathsf{Z} = \mathbf{g}^{\mathsf{z}}$. Let z be as follows:

$$
\mathsf{z} := ((\rho + u)r, r), \quad r \in_R \mathbb{Z}_q
$$

where $u \in \mathbb{Z}_q$ is system-wide unique and of the form $(c||UI)2^{l_\rho}$, and $\rho = \rho_1 2^{1+l_c+l_{UI}+l_{\rho_2}} + \rho_2$, where $\rho_1 \in_R \{0, 1\}^{l_{\rho_1}}$ and $\rho_2 \in_R \{0, 1\}^{l_{\rho_2}}$. Notice that c is a random counter and $(\rho + u) = \rho_1 ||0||c||UI||\rho_2$ holds.

The initialization process for DKG2 is the same as InitDKG1. Since only the system parameters (the bit-lengths) are slightly different, no extra protocol is given here.

Several steps of the key generation protocol DKG2 are analogous to the ones of DKG1. In step 1, the players jointly increment the counter contained in u. Moreover, they update the corresponding P-commitments. In step 2a, each player generates shares of the random padding ρ, where the upper l_{ρ_1} and the lower l_{ρ_2} bits are chosen at random with respect to normal distribution and all other bits are verifiably set to 0. This is done via JPVSS2. Then, in step 2b, the players use JPVSS to jointly generate shares

---

Protocol DKG2

Input (P_i): $([u_i]^q, [u'_i]^q)$, $([1_i]^q, [1'_i]^q)$, $\{\widetilde{U}_j, \widetilde{1}_j\}_{j \in I(\mathcal{Q})}$

Output (P_i): $\langle z_i \rangle^q$, $([u_i]^q, [u'_i]^q)$, $\{Z_j, \widetilde{U}_j\}_{j \in I(\mathcal{Q})}$, Z

1. Each P_i computes $[u_i]^q = [u_i]^q + [1_i]^q$, $[u'_i]^q = [u'_i]^q + [1'_i]^q$ and $\{\widetilde{U}_j = \widetilde{U}_j \widetilde{1}_j\}_{j \in I(\mathcal{Q})}$.

2. The players then run the following two steps in parallel:

   a) Shares of a secret padding ρ are generated via JPVSS2, where $BL = \{(1 + l_c + l_{UI} + l_{\rho_2}, l_{\rho_1}), (0, l_{\rho_2})\}$. The output of player P_i is $([\rho_i]^q, [\rho'_i]^q)$ and $\{\widetilde{\rho}_j\}_{j \in I(\mathcal{Q})}$.

   b) Shares of a random secret $r \in \mathbb{Z}_q$ are generated via JPVSS. The output of player P_i is $([r_i]^q, [r'_i]^q)$ and $\{\widetilde{R}_j\}_{j \in I(\mathcal{Q})}$.

3. The players run VMUL2 to compute the fusion-shares of $((u + \rho)r, r)$. Hereby, each P_i gives the inputs $([\rho_i]^q + [u_i]^q, [\rho'_i]^q + [u'_i]^q)$, $([r_i]^q, [r'_i]^q)$ and $\{\widetilde{U}_i \widetilde{\rho}_i, \widetilde{R}_i\}_{i \in I(\mathcal{Q})}$ and gets $([o_{f_i}]^q, [r_i]^q)$ and $\{Z_j\}_{j \in I(\mathcal{Q})}$ as output.

4. Each P_j sets his fusion-share $\langle z_j \rangle^q = ([o_{f_j}]^q, [r_j]^q)$.

5. Finally, each player computes the public key $Z = \prod_{i \in I(\mathcal{Q})} Z_i^{(\lambda_i, 0)}$.

Fig. 5.16: Shared Collision-Free Pair Generation.

of a random secret r, uniformly distributed over $\mathbb{Z}_q$. In step 3, the protocol VMUL2 is run to provide each player with a t-degree fusion-share of the secret $((\rho + u)r, r)$ and the FD-commitments of the fusion-shares of all other players. Then, each player stores his fusion-share as $\langle z_i \rangle^q$ and computes the public key Z using a hidden interpolation.

5.6.3 Correctness

To show the correctness of protocol DKG2, one has to prove that the requirements (a), (b) and (c) of R1 hold. Notice that all parties listen to a broadcast channel and hence every player determines the same set $\mathcal{Q}$. Again, it is assumed that the init-protocol was successful and that each player was provided with correct shares and commitments.

Theorem 5.6.1 *With the assumption that solving the Discrete Logarithm Problem is hard in $\mathbb{G}_q$, protocol DKG2 fulfills R1 if at most $t < n/2$ players are actively dishonest.*

Proof. It has to be shown that (a), (b) and (c) of R1 hold.

(a) In step 1, each player locally computes $[u_i]^q$ and P-commitments of all other shares. Since no communication takes place, the values are correct. In step 2, the players jointly generate the padding ρ for u and the random part r of the secret key. Since the protocols JPVSS2 and JPVSS are used, it is guaranteed that each honest player owns a share of ρ, a share of r and the P-commitments of all other shares.

Notice that actively dishonest parties are disqualified and the shares of the honest players uniquely define ρ and r. In step 3, $(\rho + u)$ and r are multiplied using the corresponding shares. Using protocol VMUL2, actively dishonest parties can be identified and disqualified, such that only honest players hold valid shares. If $t < n/2$, then the secret can be uniquely defined over $t + 1$ shares.

(b) It suffices to show that correct shares can be distinguished from incorrect ones. A fusion-share $\langle z_i \rangle^q$ is correct, if $[r_i]^q$ corresponds to $\widetilde{R}_i$, $[u_i]^q$ corresponds to $\widetilde{U}_i$ and $\langle z_i \rangle^q$ corresponds to Z_i. The first two hold with computational uniqueness, since P-commitments are computational binding. The third holds, since FD-commitments are perfect binding and the Σ-protocol ensures that Z_i is correct.

(c) Malicious parties can be uniquely identified and disqualified. Thus, their inputs and corresponding commitments can be identified and invalidated locally by every honest player. Since the public key Z is uniquely computed over the commitments of the honest players, every of these players holds the same Z.   □

5.6.4   Secrecy

Contrary to protocol DKG1, one can *prove* the secrecy of DKG2 with respect to the computational assumption that computing discrete logarithms in $\mathbb{G}_q$ is hard. This comes with the fact that z_2 is uniformly distributed over $\mathbb{Z}_q$.

Lemma 5.6.1 *Let* $y = g^{(x_1, x_2)}$. *Then, given* y *and* g, *computing* x_1 *or* x_2 *is computationally equivalent to solving the Discrete Logarithm Problem in* $\mathbb{G}_q$.

Proof. From Theorem 4.3.4 it follows, that if one can solve the Discrete Logarithm Problem, (x_1, x_2) can be obtained. For the converse the existence of an oracle $\mathcal{O}$ is assumed, which on input y and g, returns x_2 (without loosing generality). Now, let $y \in \mathbb{G}_q$ and $g \in \mathbb{G}_q \setminus \{1\}$, where one wants to find $dlog_g(y)$, i.e. wants to solve an instance of the ordinary Discrete Logarithm Problem. Querying $\mathcal{O}((1, y), (g, 1))$ gives $(0, dlog_g(y))$, since $(g, 1)^{(0, dlog_g(y))} = (g^0 1^{-dlog_g(y)}, g^{dlog_g(y)} 1^0) = (1, y)$. The proof for the computation of x_1 can be given with analogous steps.   □

Theorem 5.6.2 *With the assumption that solving the Discrete Logarithm Problem is hard in* $\mathbb{G}_q$, *protocol* DKG2 *fulfills R2 if at most* $t < n/2$ *players are actively dishonest.*

Proof. The poly-shares of u and r are hidden using P-commitments which are perfect hiding. Hence, from the commitments one cannot learn anything about u or r. An adversary, having access to t shares of u or r cannot learn anything new, since poly-sharing is used. This also holds for the case that the t random inputs for the generation of r are publicly known, since r is the modular sum of secret inputs that are uniformly distributed over $\mathbb{Z}_q$. The same argument holds for the re-sharing procedure of r and $(u + \rho)r$. Notice, however, that the adversary can learn something about ρ, since it is computed via a shared random sum which is normally distributed. So the uncertainty about ρ relies in $l_{\rho_1} + l_{\rho_2} - 2log_2(t)$ bits if t inputs are published by malicious parties. The used Σ-proof reveals no information about the proven knowledge, since it is zero-knowledge in the non-interactive setting. Finally, it needs to be shown that given Z and $\mathbf{g}$, computing $\mathbf{z} = ((u+\rho)r, r)$ is hard. From Lemma 5.6.1 it follows that computing r from Z is equivalent to solving the Discrete Logarithm Problem in $\mathbb{G}_q$. The same holds if one wants to obtain $(u + \rho)r$. $\qquad\square$

Notice that the non-uniformity of the probability distribution of the padding has no real effect on the security of the protocol. In Section 5.8, two further variants of DKG2 are sketched which provide the same level of secrecy with much more efficiency. Nevertheless, DKG2, although less efficient, provides an increased level of privacy with respect to the design principles of Chapter 3. This is discussed further in Section 5.8.1.

5.6.5 Uniqueness

Theorem 5.6.3 *With the assumption that solving the Discrete Logarithm Problem is hard in $\mathbb{G}_q$, protocol DKG2 fulfills R3 if at most $t < n/2$ players are actively dishonest.*

Proof. To show that uniqueness is fulfilled, one has to prove that the pair $((\rho + u)r, r)$ is system-wide unique. Since multiplication in $\mathbb{Z}_q$ is bijective, $((\rho + u)r, r)$ is unique if $(\rho + u)$ is unique (cf. Theorem 3.2.1). It remains to show that $(\rho + u)$ is unique. Notice that the init-protocol guarantees that $[u_i]^q$ is a poly-share of $u = (c||UI)2^{l_{\rho_2}}$, where UI is a system-wide unique identifier. In step 1, the players jointly increment u, using shared addition of u and $2^{l_{UI}+l_{\rho_2}}$. Since no communication takes place, u remains unique. In step 2a, the players jointly generate shares of ρ. Protocol JPVSS2 ensures that the random bits in ρ are restricted to the bit-positions 0 to $l_{\rho_2} - 1$ and the bit-positions $1 + l_c + l_{UI} + l_{\rho_2}$ to $l_{\rho_1} + l_c + l_{UI} + l_{\rho_2}$. Hence, computing $(\rho + u)$ results in $\rho_1||0||c||UI||\rho_2$ which still preserves uniqueness. As was the case in DKG1, the key generation protocol can be repeated 2^{l_c} times without overflow. $\qquad\square$

5.7 Distributed Key Generation Protocol DKG3

In this section, a protocol for shared collision-free number generation is given (cf. Section 5.4.4). As mentioned, the idea is that each player computes a system-wide unique prime using an ordinary collision-free number generator. The secret key $z \in \mathbb{Z}_q$ is then defined as the product of all primes, or at least of the primes of the honest players. Let l_p be the bit-length of each chosen prime. Then $l_q = 1 + nl_p$ needs to hold in order

Protocol DKG3

Input (P_i): $\{vk_j\}_{j \in I(\mathcal{P})}$

Output (P_i): $[z_i]^q$, $\{Z_j\}_{j \in I(\mathcal{Q})}$, Z

1. Each $P_i \in \mathcal{P}$ runs the following steps in the trusted environment:

   a) Compute p_i by using **CFNG**.

   b) If $p_i \notin \mathbb{P}$ then goto the previous step.

   c) Choose $p_i' \in_R \mathbb{Z}_q$ and compute $\widetilde{P}_i = g_1^{p_i} g_2^{p_i'}$.

   d) Sign $\widetilde{P}_i$ on behalf of sk_i resulting in the signature σ_i.

2. Each P_i shares p_i over all players in $\mathcal{P}$ using **PVSS**. Hereby, he uses $\widetilde{P}_i$ as the P-commitment of p_i when broadcasting the P-commitments of the sharing coefficients. Moreover, he broadcasts the signature σ_i which needs to be verified by each player. If more than t players broadcast a complaint for the failure of the signature-verification, P_i is disqualified. In the end of this variant of **PVSS** each P_j holds $\{([p_{ij}]^q, [p_{ij}']^q)\}_{i \in I(\mathcal{Q})}$ and $\{\widetilde{P}_{ik}\}_{i,k \in I(\mathcal{Q})}$.

3. The players run **VMUL** over the shares of the secret primes of all other players of $\mathcal{Q}$. In the end, each P_i holds $([z_i]^q, [z_i']^q)$ and $\{\widetilde{Z}_j\}_{j \in I(\mathcal{Q})}$.

4. Each P_i broadcasts $Z_i = g^{[z_i]^q}$ and a proof τ_i that $dlog_g(Z_i) = dlog_{g_1}(\widetilde{Z}_i g_2^{-[z_i']^q})$.

5. For each j, every player P_i checks if $V(\tau_j, (Z_j, \widetilde{Z}_j)) = 1$ and broadcasts a complaint against P_j if not. If there are more than t complaints, then P_j is disqualified.

6. Finally, the public key is computed: $Z = \prod_{i \in I(\mathcal{Q})} Z_i^{\lambda_i}$.

Fig. 5.17: Shared Multiplication of Unique Primes.

to ensure that $z < q$. Notice that one has to expect that up to t players publish their shares. Thus, the uncertainty relies on a number that is log-normally distributed over $[0, 2^{(t+1)l_p} - 1]$. In Section 5.4.4, two alternatives for the shared generation of z were suggested. Here the second one is used, requiring a trusted device for each player.

5.7.1   Shared Multiplication of Unique Primes

Each player is provided with a collision-free number generator **CFNG**, which is assumed to output numbers that are uniformly distributed over the set of primes smaller than 2^{l_p}. Notice that this is an artificial assumption (primes are actually generated by **CFNG**) but it simplifies the analysis with respect to requirement R2. During the setup of the system, each **CFNG** is initialized appropriately with respect to its specification. Thereby, a key pair (sk_i, vk_i) is assigned to each player P_i, where sk_i is a signature key, generated and stored in a tamper-resistant environment in which CFNG is implemented. The corresponding verification key vk_i is made public.

At the beginning of protocol DKG3 (cf. Figure 5.17), each player P_i computes a number $p_i \in [0, 2^{l_p} - 1]$ in his trusted environment using CFNG. If p_i is not a prime, then the process is repeated, otherwise a P-commitment of p_i is generated and signed by the trusted environment. The signature is necessary to ensure that p_i is actually a system-wide unique prime, i.e. has been generated by the CFNG. In step 2, p_i is shared over all the other players. Although p_i is restricted to certain bit-position, there is no necessity to prove this fact. In PVSS, the dealer (here P_i) has to broadcast commitments of the sharing-coefficients. In the current situation it suffices to use the P-commitment $\widetilde{P}_i$ for the shared prime. All other commitments are generated with respect to PVSS. Furthermore, the signature σ_i must be broadcasted along with the P-commitments. Accordingly, the authenticity of $\widetilde{P}_i$ needs to be verified by each receiver using the corresponding authentic verification key vk_i. Shares of primes that do not come from a registered tamper-resistant device are rejected and the corresponding player disqualified. All other steps of PVSS are performed as stated in Section 5.2.4.2. In step 3 of protocol DKG3, the players in $\mathcal{Q}$ run VMUL over all shared primes of the honest players. Finally, the players jointly compute a D-commitment of z. Therefore, each P_i broadcasts the D-commitment Z_i of his share and proves in zero-knowledge, that it corresponds to the P-commitment $\widetilde{Z}_i$ possessed by every other player. Therefore, he uses the same non-interactive Σ-proof τ_i, as described at the end of Section 5.5.2. Finally, the public key Z is generated using a hidden interpolation of z.

5.7.2 Correctness

Theorem 5.7.1 *With the assumption that solving the Discrete Logarithm Problem is hard in $\mathbb{G}_q$, protocol DKG3 fulfills R1 if at most $t < n/2$ players are actively dishonest.*

Proof. It has to be shown that (a), (b) and (c) of R1 hold.

(a) The secret z is the product of at least $t + 1$ system-wide unique primes. Each prime has been shared among the players using PVSS. The standard verifiable multiplication protocol is run only over the correctly distributed shares where the values submitted by dishonest players are uniquely identified and eliminated. Hence, at the end of step 3, each player has a correct share of the same secret z.

(b) It must be shown that correct shares can be distinguished from incorrect ones. This can be done since steps 3-5 ensure that each accepted $[z_i]^q$ corresponds to Z_i.

(c) At the end of step 5, every player knows which D-commitment Z_i is correct. Over these commitments, he can locally compute Z (cf. step 6). $\qquad\square$

5.7.3   Secrecy

Similarly to the analysis stated in Section 5.5.4, a security analysis is sketched in place
of a formal proof. To ensure uniqueness it must be guaranteed that the product of n
primes (that are formed according to the specification) is not reduced modulo q, i.e.
$z < q$ holds. Since at least $t + 1$ player are honest, z contains at least $t + 1$ prime-
factors, each of which is uniformly distributed over the set of primes smaller than 2^{l_p}.
It is assumed that the primes of the dishonest players are known by the adversary,
although this is an artificial assumption because the primes are generated in a trusted
environment. The primes of the honest players are completely unknown because only
P-commitments are known to the adversary. Hence, the uncertainty about z concerns
a number $\hat{z} = \frac{z}{\prod_{i \in I(\mathcal{P} \setminus \mathcal{Q})} z_i}$ that lies in the interval $[0, 2^{(t+1)l_p} - 1]$ and is log-normally
distributed. The following three situations arise:

1. *Interval Discrete Logarithm Problem:* The attacker applies Pollard's kangaroo
   method to obtain $\hat{z}$ from $\widehat{Z} = g^{\hat{z}}$. Hence, the expected running time is $O(\sqrt{2^{(t+1)l_p}})$.

2. *Expected Value Discrete Logarithm Problem:* For log-normally distributed numbers
   the expected value (in the continuous case) is $E(X) = e^{\mu + \sigma^2/2}$, where μ is the mean
   and σ the standard variant of the logarithm of the variable X corresponding to $\hat{z}$.
   So the expected running time is $O(\sqrt{e^{\mu + \sigma^2/2}})$.

3. *Probability Distribution Discrete Logarithm Problem:* Since the product of uniformly distributed numbers approaches log-normal distribution, the expected running time of the baby-giant step algorithms is $O(\sqrt{E(X)})$.

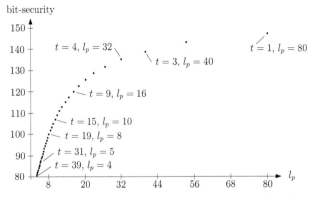

Fig. 5.18: Bit-Security for Various Settings for t and l_p where $(t + 1)l_p = 160$.

A further attempt for an adversary to obtain z could take advantage of the fact that z is the product of at least $t+1$ primes, each of which is l_p bits long. From Theorem 5.2.2, the number of primes smaller than 2^{l_p} is around $\frac{2^{l_p}}{l_p}$. In order to find $\hat{z}$, the adversary has to compute all possible products of $t+1$ primes of length l_p. Therefore, he approximately has to compute $(\frac{2^{l_p}}{l_p})^{t+1}$ candidates for $\hat{z}$. Finding $\hat{z}$ using brute force should not be more efficient than solving the Interval Discrete Logarithm Problem. For the running time of Pollard's kangaroo algorithm $c\sqrt{2^{l_p(t+1)}}$, c a constant, one observes

$$log_2(c\sqrt{2^{l_p(t+1)}}) = log_2(c) + \frac{1}{2}(t+1)l_p$$

On the contrary, for the brute force attack one approximately gets

$$log_2((\frac{2^{l_p}}{l_p})^{t+1}) = l_p(t+1) - log_2(l_p)(t+1)$$

Based on these two estimations, one has to find the breakover point, where each l_p is sufficiently long to make a brute force attack approximately as hard as solving the Interval Discrete Logarithm Problem. Figure 5.18 shows the bit-security for different settings for which the lower security bound is assumed to be $l_p(t+1) = 160$. If $l_p = 4$, $t = 39$ and the bit-security is 80. Increasing the number of bits of each prime decreases the number of possible (honest) players exponentially, but at the same time increases the bit-security. For instance, having $l_p = 16$, we get $t = 9$ and a bit-security of 120. The lower bound for n is 3, since for correctness $t < n/2$ must hold. Then, $t = 1$ and $l_p = 80$ and the bit-security is maximal, i.e. around 147 bits. Notice, however, that these are just brief estimations and that the bit-security only holds for a brute-force attack. Regarding Pollard's rho or kangaroo algorithm, a maximal bit security of $t_p(t+1)/2 = 80$ is achieved for this example.

5.7.4 Uniqueness

Theorem 5.7.2 *With the assumption that solving the Discrete Logarithm Problem is hard in $\mathbb{G}_q$, protocol DKG3 fulfills R3 if at most $t < n/2$ players are actively dishonest.*

Proof. From the Fundamental Theorem of Arithmetic (cf. Theorem 5.2.1), each natural number can be represented as the product of a unique sequence of primes. From steps 1 and 2 it is guaranteed that each *accepted* prime is chosen by a collision-free number generator *and* is restricted to the interval $[0, 2^{l_p} - 1]$. Since $l_q = 1 + nl_p$, $z < q$ holds and z is system-wide unique. $\square$

5.8　Optimization of DKG2

In Section 5.6.4 it has been shown that given $\mathbf{Z} = \mathbf{g}^{(z_1, z_2)}$, computing z_1 or z_2 is computationally equivalent to solving the Discrete Logarithm Problem. Thus, the use of setting $\mathbf{z} = (ur, r)$ is questionable since the secrecy in any case fully relies on r. In this section, the variants where $\mathbf{z} = (ur, r)$ and $\mathbf{z} = (u, r)$ are compared with respect to the design principles of collision-free number generators, the requirement for secrecy and efficiency. Furthermore, two approaches are sketched for the shared generation of $\mathbf{z} = (u, r)$ and the corresponding public key $\mathbf{Z}$. The first variant is similar to DKG2 and requires the involvement of Σ-proofs. The second variant is similar to DKG (cf. Section 5.2.6.1) and gets by with homomorphic commitments exclusively.

5.8.1　Comparison

An optional requirement in the design of a collision-free number generator is the unlinkability of numbers, that have been generated by the same instance (cf. Section 3.1.2). Now consider two secret keys $\mathbf{z} = (u, r)$ and $\mathbf{z}' = (u', r')$, where $r = r'$ holds. Since $u \neq u'$, $\mathbf{z} \neq \mathbf{z}'$ holds. However, one is able to decide that both numbers have been generated by the same set of players: let $(A, B) = \mathbf{g}^{(u,r)} = (g_1^u g_2^{-r}, g_1^r g_2^u)$ and $(C, D) = \mathbf{g}^{(u',r)} = (g_1^{u'} g_2^{-r}, g_1^r g_2^{u'})$. Then one can compute:

$$E = AC^{-1} = g_1^u g_2^{-r} (g_1^{u'} g_2^{-r})^{-1} = g_1^{u-u'} g_2^{-r+r} = g_1^{u-u'}$$
$$F = BD^{-1} = g_1^r g_2^u (g_1^r g_2^{u'})^{-1} = g_1^{r-r} g_2^{u-u'} = g_2^{u-u'}$$

If u and u' have been derived from the same identifier and the padding used is small or no padding is used, then one can run a brute force attack to decide if $dlog_{g_1}(E) = dlog_{g_2}(F)$. Hence, linkability is vulnerable if $r = r'$.

Now consider the case where $\mathbf{z} = (ur, r)$ and $\mathbf{z}' = (u'r, r)$. Then $(A, B) = (g_1^{ur} g_2^{-r}, g_1^r g_2^{ur})$ and $(C, D) = (g_1^{u'r} g_2^{-r}, g_1^r g_2^{u'r})$, and one can compute:

$$E = AC^{-1} = g_1^{ur} g_2^{-r} (g_1^{u'r} g_2^{-r})^{-1} = g_1^{r(u-u')} g_2^{-r+r} = g_1^{r(u-u')}$$
$$F = BD^{-1} = g_1^r g_2^{ur} (g_1^r g_2^{u'r})^{-1} = g_1^{r-r} g_2^{r(u-u')} = g_2^{r(u-u')}$$

This time, however, $(u - u')$ is blinded by r. Thus, to decide if $dlog_{g_1}(E) = dlog_{g_2}(F)$, one has to decide if $(g_1^w, g_1^{r(u-u')}, g_1^{wr(u-u')})$ is a Diffie-Hellman Triple. Since w and r are uniformly distributed over $\mathbb{Z}_q$, the Decision Diffie-Hellman Problem must be solved.

A similar comparison can be given for DKG1. Notice, however, that the situation there is much more dangerous, because setting $z = u \| r$ has the disadvantage that a sequence of generated keys can be compressed.

Although secret keys of the form (u, r) give a privacy hole in some situations, two protocols for its shared generation are given anyway as they are very efficient. Moreover, linkability is only an optional requirement.

5.8.2 Shared Generation with Σ-Proofs

The goal is to provide each player with a fusion-share of $z = (u, r)$ and the public key $Z = g^{(u,r)}$. Since the secrecy fully relies on r, the use of any padding for u is omitted. For simplicity $u := UI\|c$ with the only condition $u < q$. The init-protocol is omitted, since it can be given analogously to Init-DKG1.

Protocol DKG2a

Input (P_i): $([u_i]^q, [u_i']^q), ([1_i]^q, [1_i']^q), \{\widetilde{U}_j, \widetilde{1}_j\}_{j \in I(Q)}$

Output (P_i): $\langle z_i \rangle^q, ([u_i]^q, [u_i']^q), \{Z_j, \widetilde{U}_j\}_{j \in I(Q)}, Z$

1. Each P_i computes $[u_i]^q = [u_i]^q + [1_i]^q$, $[u_i']^q = [u_i']^q + [1_i']^q$ and $\{\widetilde{U}_j = \widetilde{U}_j\widetilde{1}_j\}_{j \in I(Q)}$.

2. Each player P_i participates in a run of JPVSS. Hereby, he gets $([r_i]^q, [r_i']^q)$ and $\{\widetilde{R}_j\}_{j \in I(Q)}$ as output (further outputs are no longer used).

3. Each P_i sets his fusion-share $\langle z_i \rangle^q = ([u_i]^q, [r_i]^q)$ and broadcasts $Z_i = g^{\langle z_i \rangle^q}$ and a non-interactive Σ-proof τ_i, that Z_i is formed correctly with respect to $\widetilde{R}_i$ and $\widetilde{U}_i$.

4. For each j, every player P_i checks if $V(\tau_j, (Z_j, \widetilde{R}_j, \widetilde{U}_j)) = 1$ and broadcasts a complaint against P_j if not. If there are more than t complaints, then P_j is disqualified.

5. Finally, each player computes the public key $Z = \prod_{i \in I(Q)} Z_i^{\langle \lambda_i, 0 \rangle}$.

Fig. 5.19: Simplified Shared Collision-Free Pair Generation with Σ-Proofs.

At the beginning of DKG2a, the players locally add the shares of u and 1. This time, the share $[1_i]^q$ is indeed a share of the value 1, rather than of $2^{l_{UI}+l_{P2}}$. Furthermore, they update the corresponding set of commitments. In step 2, an ordinary run of JPVSS is performed which provides each player with a share of a secret r, which is uniformly distributed over $\mathbb{Z}_q$. As a by-product, the players obtain P-commitments of all other shares. In step 3, each player can form his fusion-share, which consists of the share of u, computed in step 1, and the share of r, obtained in step 2. It remains to compute Z. Each P_i broadcasts an FD-commitment of his fusion-share and proves in zero-knowledge that the components of the committed pair correspond to the P-commitments $\widetilde{U}_i$ and $\widetilde{R}_i$ respectively. This requires standard techniques for proving the equality of discrete logarithms. Using an appropriate verification function V, each receiver checks if the broadcasted D-commitments are correct and finally computes the public key through step 5.

Correctness can be shown straightforwardly, since DKG2a is a simplification of DKG1 and DKG2. Secrecy still holds using Lemma 5.6.4. Uniqueness is trivial, since a pair (u, r) is always unique for a unique u.

Protocol InitDKG2b

Output (P_i): $([u_i]^q, [u'_i]^q)$, $([1_i]^q, [1'_i]^q)$, $\{[u_{ji}]^q,\ [u'_{ji}]^q,\ [1_{ji}]^q,\ [1'_{ji}]^q,\ \widetilde{U}_{ji},\ \widetilde{1}_{ji}\}_{j \in I(\mathcal{Q})}$, $\{a_{ik}, b_{ik}\}_{k \in \mathcal{T}}$

1. The initializer chooses $UI \in \{0,1\}^{l_{UI}}$ and computes $u = UI2^{l_c}$.

2. Then he chooses $\{u_i, 1_i \in_R \mathbb{Z}_q\}_{i=1}^{n-1}$ and $\{u'_i, 1'_i \in_R \mathbb{Z}_q\}_{i=1}^{n}$.

3. Next, the initializer computes $u_n = u - \sum_{i=1}^{n-1} u_i$ and $1_n = 1 - \sum_{i=1}^{n-1} 1_i$.

4. Now, he locally simulates two executions of JPVSS, where $\{(u_i, u'_i)\}_{i=1}^{n}$ and $\{(1_i, 1'_i)\}_{i=1}^{n}$ serve as the secret inputs and blindings for the P-commitments, respectively. The (necessary) output for player P_i is then $([u_i]^q, [u'_i]^q)$, $([1_i]^q, [1'_i]^q)$, $\{[u_{ji}]^q$, $[u'_{ji}]^q, [1_{ji}]^q, [1'_{ji}]^q, \widetilde{U}_{ji}, \widetilde{1}_{ji}\}_{j \in I(\mathcal{Q})}$, and the set of sharing-coefficients $\{a_{ik}, b_{ik}\}_{k \in \mathcal{T}}$.

Protocol DKG2b

Input (P_i): $([u_i]^q, [u'_i]^q)$, $([1_i]^q, [1'_i]^q)$, $\{[u_{ji}]^q,\ [u'_{ji}]^q,\ [1_{ji}]^q,\ [1'_{ji}]^q,\ \widetilde{U}_{ji},\ \widetilde{1}_{ji}\}_{j \in I(\mathcal{Q})}$, $\{a_{ik}, b_{ik}\}_{k \in \mathcal{T}}$

Output (P_i): $\langle z_i \rangle^q$, $([u_i]^q, [u'_i]^q)$, $\{[u_{ji}]^q, [u'_{ji}]^q\}_{j \in I(\mathcal{Q})}$, $\{a_{ik}\}_{k \in \mathcal{T}}$, Z

1. Each P_i computes $[u_i]^q = [u_i]^q + [1_i]^q$, $[u'_i]^q = [u'_i]^q + [1'_i]^q$, $\{[u_{ji}]^q = [u_{ji}]^q + [1_{ji}]^q$, $[u'_{ji}]^q = [u'_{ji}]^q + [1'_{ji}]^q$, $\widetilde{U}_{ji} = \widetilde{U}_{ji}\widetilde{1}_{ji}\}_{j \in I(\mathcal{Q})}$ and $\{a_{ik} = a_{ik} + b_{ik}\}_{k \in \mathcal{T}}$.

2. Each player P_i participates in a run of JPVSS. Hereby, he gets $([r_i]^q, [r'_i]^q)$, $\{([r_{ji}]^q, [r'_{ji}]^q)\}_{j \in I(\mathcal{Q})}$ and $\{c_{ik}\}_{k \in \mathcal{T}}$ as output. The commitments can be removed.

3. Each P_i sets $\langle z_i \rangle^q = ([u_i]^q, [r_i]^q)$ and broadcasts $\{D_{ik} = g^{(a_{ik}, c_{ik})}\}_{k \in \mathcal{T}}$.

4. Each P_j computes $\{Z_{ij} = g^{([u_{ij}]^q, [r_{ij}]^q)}\}_{i \in I(\mathcal{Q})}$ and verifies if $Z_{ij} = \prod_{k \in \mathcal{T}} D_{ik}^{(j^k, 0)}$ holds for all $i \in I(\mathcal{Q})$. If the check fails for any i, P_j complains against P_i by broadcasting $([u_{ij}]^q, [u'_{ij}]^q)$ and $([r_{ij}]^q, [r'_{ij}]^q)$ which satisfy $\widetilde{U}_{ji}$ and the verification in JPVSS, but not the one above. Each P_i participates in the reconstruction phase of u_j and r_j for every player P_j, against whom a valid complaint has been broadcasted.

5. Finally, each player computes the public key $Z = \prod_{j \in I(\mathcal{Q})} D_{j0}$.

Fig. 5.20: Simplified Shared Collision-Free Pair Generation without Σ-Proofs.

5.8.3   Shared Generation without Σ-Proofs

Interestingly, if one modifies the init-protocol appropriately, a key generation protocol can be given without Σ-proofs. The idea is to give a similar solution as used in protocol

DKG. If the same techniques used there are applied, then an init-protocol must be designed to provide each player with shares and poly-coefficients of u, analogous to the distributed generation of r. In DKG, each player stores the poly-coefficients of his own secret input, that he gives as input to JPVSS. This enables and efficient computation of the D-commitments of all key-shares and of all shares of the secret inputs. In the current situation this idea is extended to take u into account. Notice that u is not generated in a distributed way so that the players do not have shares of secret inputs or poly-coefficients, but rather only poly-shares of u. Thus, the idea is that during protocol Init-DKG2b, the initializer locally computes n sum-shares of u and 1 in $\mathbb{Z}_q$, and chooses n random sum-shares, which uniquely define the random secret blindings u' and $1'$. In the next step he locally simulates JPVSS over all sum-shares of u and u' and over all sum-shares of 1 and $1'$. Here, (u_i, u_i') and $(1_i, 1_i')$ are interpreted as the secret inputs of P_i in a real protocol-run. Since all computations are carried out locally, the initializer does not need to perform any verifications. In the end each player P_i is provided with his poly-shares of u, u', 1 and $1'$. Moreover, he has n poly-shares of each *sum-share*, the corresponding P-commitments and the coefficients that have been used to share the sum-shares u_i and 1_i. Using this initialization, every player holds all values which he would have received if u had been generated in a shared way.

At the beginning of protocol DKG2b, each player P_i locally adds his poly-shares of u and 1. Moreover, he updates the poly-shares of the corresponding blinding-values. The same updates are necessary for the shares of all other sum-shares and the P-commitments. Finally, he updates his own poly-coefficients with respect to u_i. In step 2, the players perform an ordinary run of JPVSS, where each player gets a share of r as output and the shares of the random inputs of all other players. Furthermore, he stores his own sharing-coefficients. Then, in step 3, he is able to form his own fusion-share of the secret key and broadcasts a set which contains the D-commitments of the coefficient-pairs $\{(a_{ik}, c_{ik})\}_{k \in \mathcal{T}}$. Based on this set, every receiver locally computes the D-commitments of all fusion-shares that correspond to the secret inputs for the generation of u and r. The remainder of the protocol works analogously to protocol DKG, so that finally each player gets Z as output.

Correctness, secrecy and uniqueness can be shown straightforwardly.

5.9   Concluding Remarks

This chapter presented several initial approaches for *collision-free* distributed key generation for discrete-logarithm-based cryptosystems. It remains to compare them to DKG, a protocol widely used in practice. An important goal in the design of multi-party protocols is to keep the number of broadcasts low and to avoid Σ-proofs.

Protocol	# BCs	Σ-Proof	Group	Security	Uniqueness
DKG	2	no	$\mathbb{G}_q$	provable	no
DKG1	3	yes	$\mathbb{G}_q$	heuristic	yes
DKG2	2	yes	$\mathbb{G}_p$	provable	yes
DKG3	$2 + \log_2(n)$	yes	$\mathbb{G}_q$	heuristic	yes
DKG2a	2	yes	$\mathbb{G}_p$	provable	yes
DKG2b	2	no	$\mathbb{G}_p$	provable	yes

Tab. 5.1: Comparison of the Protocols.

As far as the number of (minimal) broadcasts (BCs) per player is concerned, protocols
DKG2, DKG2a and DKG2b are the only ones which are as efficient as DKG. Additionally, DKG2b is the only one that works without Σ-proofs. The security of these
protocols can be reduced to the Discrete Logarithm Problem in $\mathbb{G}_q$, but the subsequent
cryptosystem then needs to be based on $\mathbb{G}_p$. The latter can be a drawback regarding
the computational costs. The security of DKG1 and DKG3 is more or less heuristic,
since the generated secret keys follow an uncommon probability distribution. Moreover, they require more broadcasts than the other protocols and require the use of
Σ-proofs. Hence, DKG1 and DKG3 are more of theoretical interest.

An open problem is to design a protocol that is provably secure while *not* being based
on the concept of fusion. Moreover, an interesting open task would be to design a
protocol for the shared generation of system-wide unique RSA moduli and the corresponding key pairs. Moreover, easier solutions could be found for the case where
keys are only generated in a distributed way and then reconstructed immediately,
i.e. not used in shared form. This would be interesting for sharing a trust center,
where security against passive adversaries might be sufficient.

Unlinkable Anonymous Authentication

Chapter 6

6.1 Introduction

During an authentication process an instance (the prover) proves his identity to another instance (the verifier). In several situations the prover only wants to convince the verifier that he is a legitimate user without disclosing any personal information. Here, there are two different point of views:

1. The *prover* wants to remain anonymous, thus at most wants to prove that he is registered, but does not want to give any further information to the verifier.

2. The *verifier* wants to know with whom he communicates, especially for the case that the prover behaves dishonestly in the protocol steps *after* the authentication.

It seems as if both views contradict each other: the prover wants to remain anonymous, whereas the verifier wants to know the identity of the prover. Standard authentication protocols have the following drawbacks concerning the privacy of the prover:

- The *identity* of the prover is (publicly) *known*.

- All authentication processes of the same prover are mutually *linkable*.

The two properties generally come with the fact that standard public-key certificates are involved, which contain the identity of the user. Obviously, a compromise between prover and verifier has to be found, such that the interests of both are preserved: maximum privacy for honest provers, and maximum security for verifiers against dishonest provers. This chapter distinguishes between three classes of *unlinkable anonymous authentication* with respect to anonymity revocation and linkability:

1. Anonymity revocation and linking are both infeasible.

2. Anonymity revocation is feasible by a trusted party, but linking is infeasible.

3. Anonymity revocation and linking are feasible by a trusted party.

The first class is the most interesting one for the prover, since the verifier has no (practical) chance to reveal his identity or mutually link any authentication processes that are run by the same prover. An example for this class is patent search or storing incremental medical records in centralized databases.

The second class also protects the privacy of the prover, but provides optional anonymity revocation by a trusted party, in case he behaves dishonestly after the authentication process. Since linking is infeasible, no other verifier can find out if he communicates to the same (dishonest) prover. Such a scheme is useful for applications where the dishonesty of a prover is only the business of the betrayed verifier. For example, consider a scenario where an instance wins an auction, but refuses to pay. Then the verifier has a strong interest in disclosing the identity of the prover. However, there is no necessity to inform other auction chairs.

The third class additionally provides optional linkability through a trusted party. An application where this could be useful is online gambling. In real life, cheating gamblers are sometimes banned from casinos nationwide. If a player behaves dishonestly in an online game, he can be identified and all other casinos can be informed.

6.1.1   Requirements

For authentication schemes, with a special focus on preserving the privacy of the prover, the following requirements naturally come to mind:

R1 (Unforgeability): A poly-bounded algorithm is not able to efficiently forge an authentication process.

R2 (Anonymity): Given a protocol transcript, a poly-bounded algorithm is not able to efficiently identify the corresponding user.

R3 (Unlinkability): Given a set of protocol transcripts, a poly-bounded algorithm is not able to efficiently decide, which of them correspond to the same prover.

R4 (Optional Anonymity Revocation and Linkability): Given a protocol transcript, a trusted revocation center is able to efficiently disclose the identity of the user or even identify all authentication processes in which he participated.

6.1.2   Related Work

Several solutions have been proposed in this area. Many of them are based on group signatures, which allow users to prove their membership of a group without revealing their identity [CvH91, CS97, ACJT00, BSZ04]. Others are based on threshold privacy where a user remains anonymous when accessing a service up to a limited number of times [TFS04, NSN05]. Revocation of anonymity and linkability are a requirement in anonymous credential systems [CL01, PM04] or electronic money [JY96]. A solution optimized for power-limited devices has been proposed in [KCKB03].

The scheme presented here is neither based on group signatures nor on threshold privacy (as described in [SYT05]). Compared to more general solutions, such as traceable signatures [KTY04], the approaches presented in this chapter are more specific. The focus lies on applications which are based on transaction-pseudonyms whilst also providing an efficient revocation and linking process. Additionally, one of the proposed schemes in this chapter is optimized for power-limited devices (e.g. smartcards).

One approach to settle the demands of the prover and the verifier is the use of digital pseudonyms, of which there are several types [PK01]: person-pseudonyms for instance, have the property that a user always uses the same pseudonym for every transaction. Another option is to provide a user with role-pseudonyms, which gives a better anonymity, but requires the establishment of more pseudonyms. Transaction-pseudonyms are the extreme case where the user is provided with a new pseudonym for every communication process. Such pseudonyms have the advantage, that the degree of mutual linking is minimal. The drawback is an increasing number of pseudonyms and hence an increasing amount of communication for their establishment.

When using a pseudonym, it must be guaranteed that it is authentic and that the holder can prove ownership. For optional anonymity revocation and linking, additional mechanisms are generally necessary. Furthermore, it must be ensured that revoking a pseudonym leads to the *unique* identification of the corresponding user.

6.1.3   Contribution and Organization

This chapter presents two mutually related authentication schemes that provably fulfil the requirements R1-R4. Prior to describing the proposals, some preliminaries are given in Section 6.2. To contrast the ideas and different properties of the two schemes, a high-level description is given in Section 6.3. Section 6.4 describes the scheme published in [SS06]. In addition to that presented in [SS06], formal security proofs are given. An improved solution is introduced in Section 6.5, that offers increased efficiency and more advanced linking-techniques as a by-product. A preliminary version of this second scheme is published in [RSS06a]. The chapter closes with a discussion of pooling pseudonyms for optional revocation and some applications.

6.2   Preliminaries

This chapter assumes that the reader is familiar with the preliminaries given in Sections 2.2, 4.2 and 5.2, respectively. Additionally techniques are given in the following.

6.2.1   Ideal Collision-Free Number Generator

A collision-free number generator is used to guarantee that a generated pseudonym is system-wide unique. For the proposed authentication schemes there is no necessity to give internal details about the used collision-free number generator and using it as a black-box is sufficient. Furthermore, the security proofs assume that the involved secrets are indistinguishable from true random numbers. So, for simplicity, it is assumed that the used collision-free number generator is *ideal*. This enables security proofs in some kind of random oracle model. Since every generated number needs to be system-wide unique, the random oracle model is not referred to in the security proofs, because this seems to give a contradiction. The ideal collision-free number generator is given through the following abstract function:

$$\mathsf{iCFNG} : \emptyset \to \mathbb{Z}_p, \quad \mathsf{iCFNG}() = o$$

Notice that instancing iCFNG with a practical solution requires a detailed analysis, which is not given here because it would go beyond the scope of this chapter.

6.2.2   The Diffie-Hellman Triple

The Decision Diffie-Hellman Problem has been defined in Section 4.2.3. Given a triple $(X, Y, Z) \in \mathbb{G}_q$ and $g \in \mathbb{G}_q$, the problem is to decide if $dlog_g(Z) = dlog_g(X)dlog_g(Y)$ holds assuming that these discrete logarithms exist. In this chapter, a triple of this form in $\mathbb{G}_q$ or $\mathbb{G}_p$[1] is referred to as the *Diffie-Hellman Triple*.

6.2.3   The Double Discrete Logarithm Problem

The Discrete Logarithm Problem is the problem of inverting the exponentiation in particular groups. The Double Discrete Logarithm Problem [Sta96] has a similar meaning, namely, that it is the problem of inverting double exponentiation. Prior to defining this problem, the meaning of a double exponentiation must be explained in this context. Let $\mathbb{G}_p$ be a group of prime order p and $\mathbb{G}_q < \mathbb{Z}_p^*$ with prime order q, where $p = 2q + 1$. For both groups it is assumed that computing discrete logarithms

---

[1]Not to confuse with the fusion-setting introduced in Chapter 4.

is hard. Furthermore, let $g \in \mathbb{G}_p \setminus \{1\}$ and $\gamma \in \mathbb{G}_q \setminus \{1\}$. The double exponentiation function is defined as follows [Sta96]:

$$\mathbb{Z}_q \rightarrow \mathbb{G}_p : \; x \mapsto g^{\gamma^x}$$

Conversely, the double discrete logarithm of $y \in \mathbb{G}_p$ to the bases g and γ is the unique element $x \in \mathbb{Z}_q$, such that $y = g^{\gamma^x}$ holds. This leads to the following definition:

Definition 6.2.1 Let $\mathbb{G}_p$ be a group and $\mathbb{G}_q < \mathbb{Z}_p^*$, where p and q are odd primes s.t. $p = 2q + 1$. Let $g \in \mathbb{G}_p \setminus \{1\}$, $\gamma \in \mathbb{G}_q \setminus \{1\}$ and $y = g^{\gamma^x}$, where $x \in \mathbb{Z}_q$. The *Double Discrete Logarithm Problem* (DDLP) is the following: given y, g, γ, p and q, find x.

6.2.4 Verifiable Encryption of a Discrete Logarithm

Consider an infrastructure where users generate their secret ElGamal keys on their own. In case of emergency or if a user does something illegal it might be necessary that a revocation center is able to recover the secret key. Hence, techniques are required by which secret ElGamal keys (i.e. discrete logarithms) can be escrowed in an encrypted way. To avoid the problem that a user escrows something other than the secret key, the correctness of the escrow process must be publicly verifiable.

In [Sta96], Stadler proposed a solution to this problem for the case that the ElGamal encryption scheme is used to verifiably escrow the inverse of a discrete logarithm. His approach is based on the Double Discrete Logarithm Problem, as defined in the previous section. Proving the knowledge of a discrete logarithm is quite simple [CEG87, Sch89] and a Σ-proof for the knowledge of the double discrete logarithm of $y = g^{\gamma^x}$ can be designed similarly: let $(t, c, s) = (g^{\gamma^\alpha}, c, \alpha - cx)$ be the protocol transcript, with $\alpha, c \in_R \mathbb{Z}_q$. The verifier would then want to check if $dlog_g(t) \equiv (dlog_g(y))^c \gamma^s \pmod{p}$. For $c \in \mathbb{Z}_q$, the verifier would have to be able to compute $g^{\gamma^{xc}}$ from y and c. Unfortunately, this is infeasible for a large c. However, a feasible solution exists if $c \in \{0, 1\}$. The verification is then two-fold:

$$t \stackrel{?}{=} \begin{cases} g^{\gamma^s} & \text{if } c = 0 \\ y^{\gamma^s} & \text{if } c = 1 \end{cases}$$

This Σ-proof, together with a Σ-proof for the knowledge of a discrete logarithm, can be used to verifiably escrow the inverse of a discrete logarithm. Let the system parameters be those used in Definition 6.2.1. Furthermore, let $h_R = \gamma^{z_R}$ be the ElGamal encryption key of a revocation center R, where $z_R \in \mathbb{Z}_q$ is the corresponding secret key and let $y = g^x$, where $x \in \mathbb{Z}_p$. The goal is that sufficient information about x is escrowed in an encrypted form, such that R can decrypt it if pre-defined conditions are met. Without loosing generality, let the prover be the holder of x and R be the verifier. The

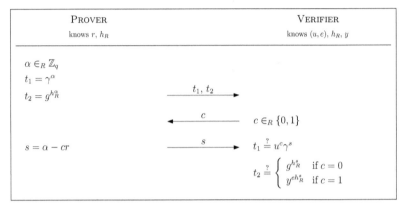

Fig. 6.1: Σ-proof, that (u, e) is an ElGamal Ciphertext of the Plaintext $(dlog_g(y))^{-1}$.

idea now is that an ElGamal ciphertext-pair (u, e) is computed such that y^e results in $g^{h_R^r}$, where $r \in_R \mathbb{Z}_q$, which can be used as a basis for a Σ-proof for the knowledge of a double discrete logarithm. Hence, the prover needs to send $(u, e) = (\gamma^r, x^{-1}h_R^r)$ to the verifier. Then, a Σ-proof for the knowledge of a discrete logarithm and a double discrete logarithm are run in parallel based on the *same* challenge $c \in \{0, 1\}$: the prover selects $\alpha \in_R \mathbb{Z}_q$ and sends $(t_1, t_2) = (\gamma^\alpha, g^{h_R^\alpha})$ to the verifier. The verifier then sends the challenge $c \in_R \{0, 1\}$ to the prover, who returns the response $s = \alpha - cr$. The verifier accepts the proof, if for k runs of the protocol, the verifications stated in the bottom of Figure 6.1 hold.

The security of this scheme relies on the Discrete Logarithm Problem (break commitments of the form g^x), the Double Discrete Logarithm Problem (break commitments of the form g^{γ^x}) and the security of the ElGamal encryption scheme. For security proofs the reader is referred to [Sta96].

6.2.5 Transforming Σ-Proofs into Zero-Knowledge Proofs

The proofs of ownership used in this chapter are Σ-proofs (cf. Section 5.2.3). Such proofs are only honest-verifier zero-knowledge so that they are generally not useful for real-world applications. In Chapter 5 such proofs were used in a non-interactive form which by-passed the problem. However, for the authentication schemes presented in the current chapter, non-interactive variants are not useful, since a real interaction is indispensable. Fortunately, a technique [Dam00] exists for transforming Σ-protocols so that they are zero-knowledge even if the verifier is dishonest. Moreover, the protocols can be run in parallel while still remaining zero-knowledge, i.e. they are generally much more efficient than those in the ordinary setting. Such transformed protocols are then called *concurrent zero-knowledge* [DNS98]. It is henceforth assumed that all protocols are variants transformed according to [Dam00]. For simplicity they are denoted in their original form.

Remark 6.2.1 Notice that an implementation of the authentication schemes given in this chapter would implicitly require the application of Damgård's techniques.

6.2.6 Mix-Nets

Consider the following problem. A user wants to send a message m to a server anonymously. Therefore, he encrypts m using the public key of the server and sends the resulting ciphertext to the server. Network protocols, such as the Internet Protocol (IP), do not provide sender-receiver-anonymity, because the IP-addresses of both are contained in every transmitted packet.

A solution to this problem is the use of mix-nets, first proposed by Chaum in [Cha81]. A typical (modern) mix-net used in a small network can be sketched as follows. Each sender encrypts his message with a public key, whose corresponding secret key is shared among a set of n mixers (the components of a mix-net) and posts it on a bulletin board. Let $\rho' = (\rho'_0, \ldots, \rho'_{k-1})$ be a vector containing all posted ciphertexts. Then every mixer has the following tasks:

1. Retrieval of ρ' from the bulletin board.

2. Re-randomization of every $\rho'_i \in \rho'$.

3. Random selection of a permutation $s : [0, k - 1] \to [0, k - 1]$.

4. Setting $\rho' := (\rho'_{s(0)}, \ldots, \rho'_{s(k-1)})$ at the bulletin board.

Typically, the transformations are done by mixer M_1 up to mixer M_n. Hereby, it is sometimes expected that a certain subset of mixers is at most passively or actively corrupt with respect to a threshold cryptosystem. After the last mixer has performed his modifications, the content of ρ' is jointly decrypted by the mixers resulting in ρ. Generally, the messages are dedicated to a particular receiver. If the receiver does not need to be anonymous, then the sender can place the IP-address of the receiver in the ciphertext. If the message is confidential then it needs to be encrypted first using the public key of the receiver and then using the public key of the mix-net.

Approaches based on the concept of re-randomization require that the applied encryption scheme provides homomorphic properties. Such encryption schemes are malleable [DDN91], which is a property that has been identified to be quite dangerous, because it enables successful chosen-ciphertext attacks. Nonetheless, for some particular applications, such as electronic voting schemes based on homomorphic encryption, it is a necessary property. A useful malleable scheme that provides re-randomization is the ElGamal encryption scheme. Let $(u, e) \in \mathbb{G}_p$ be an ElGamal ciphertext with respect to the public key $h \in \mathbb{G}_p$, the generator g, and the plaintext $m \in \mathbb{G}_p$. Then (u, e) can be re-randomized into a new ciphertext (u', e') as follows:

$$(u', e') = (u, e)(g^s, h^s), \quad s \in_R \mathbb{Z}_p$$

where multiplication is applied component-wise in $\mathbb{G}_p$. Since group exponentiation forms a homomorphism, (u', e') is a valid ElGamal ciphertext:

$$(u', e') = (u, e)(g^s, h^s) = (ug^s, eh^s) = (g^r g^s, mh^r h^s) = (g^{r+s}, mh^{r+s}) = (g^{r'}, mh^{r'})$$

For applications where the public key needs to be unknown during a re-randomization, universal re-randomization is a useful technique [GJJS04]. Furthermore, to provide security against active adversaries requires that each mixer can prove that the new vector ρ' posted at the bulletin board (known as verifiable shuffling), contains the same encrypted plaintexts as the former ones [Hir01]. The challenge here is to provide efficient zero-knowledge proofs which do not reveal any relation between the ciphertexts in the previous version of ρ' and the ciphertexts in the current version of ρ'.

For Internet applications, a practical mixing technology is Onion-Routing which combines mixing with ordinary routing [GRS96, RSG96].

6.3 High-Level Description

The properties of the two authentication schemes proposed in this chapter are quite similar concerning the requirements stated in Section 6.1.1. Hence, the current section gives an overview of the common ideas behind the schemes. This includes the identifier-based paradigm for the generation of pseudonyms, the general system architecture and the stages through which the two schemes run. Finally, an abstract view on the differences between the schemes is presented, followed by a check-list which acts as a guide to prove the requirements R1-R3 later on.

6.3.1 Identifier-Based Approach for Optional Revocation

A very strong level of anonymity and unlinkability can be achieved if the user generates a new random pseudonym (transaction-pseudonym) every time he participates in a communication process. This approach, however, only preserves the interests of the user but not the ones of the verifier. To overcome this drawback, a form of hidden linking between a transaction-pseudonym and the corresponding user is used which can be disclosed in case of emergency or if the user misbehaves later on. For efficiency it makes sense to use an identifier-based approach. All pseudonyms of the same user are derived from an identifier which is uniquely assigned to him. Obviously, it is intrinsic that the derivation process is one-way, such that given a pseudonym, a poly-bounded adversary is not able to efficiently obtain the corresponding user-identifier. Nevertheless, in some cases a trapdoor needs to exist, such that the used identifier can *optionally* be disclosed by a revocation center. To provide a very simple concept of anonymity revocation and linking, public and private user-identifiers are defined here:

1. The *public user-identifier* ID is known by a trust center (or is even public) and is used to derive the pseudonyms of the user.

2. The *private user-identifier* id is only known by the user himself or shared among a set of trusted instances. It uniquely corresponds to the public user-identifier.

ID is derived from id with negligible costs, but given ID, finding id is required to be infeasible for a poly-bounded algorithm. This leads to a two-stage revocation concept:

1. *Anonymity Revocation:* Given a pseudonym η, a trusted party can determine the corresponding public user-identifier ID and thus identify the user. ID, however, is not sufficient to find other pseudonyms derived from ID.

2. *Linking:* If id can be disclosed (or at least used) by a trusted party, then (a subset of) all pseudonyms derived from ID can be found with negligible costs.

Before an authentication process can take place, the verifier must be convinced that a pseudonym on hand is authentic. This can be achieved if the pseudonym has been signed by a trusted party. Hereby, it is important that such a party does not store any linking information between the signature and the user-request or the pseudonym. Such a property can be achieved by performing computations within a tamper-resistant device or by use of a blind signature scheme [Cha81, HMP94, CPS95].

Once the verifier is convinced that the pseudonym is authentic, the user has to prove the ownership of the pseudonym. This is achieved using transformed Σ-proofs.

For an optional revocation process, trapdoor information about the pseudonym generation process must be escrowed at a revocation center or contained in the pseudonym itself in hidden form. To provide maximum privacy, the tasks of the revocation center should be shard among a set of instances.

6.3.2 Guaranteed Unique Identification

To provide unique identification of a user, each private user-identifier is generated by a collision-free number generator. The generation process can be run by the user himself (assuming he is honest or in possession of a smartcard) or by a trusted party during the user's registration. For a unique correspondence between the public and the private user-identifiers and to fulfill the desired properties, an injective one-way function is used to compute ID from id. Additionally, the anonymity revocation process *must* lead to a unique identification of the user. In principle, the generation and revocation of a pseudonym can be seen as a commitment scheme (cf. Section 5.2.2). The commitment (the pseudonym) must be at least computational binding and hiding: on the one hand the revocation process must uniquely lead to the user-identifier (binding) and on the other hand it must be infeasible to extract the user-identifier from the pseudonym before the revocation process takes place (hiding).

6.3.3 System Architecture

The system consists of users, a pseudonym certification authority (PCA), a bulletin board (BB) and a set $\mathcal{R}$ of revocation centers $RC_1, \ldots, RC_n$. In the proposed authentication protocols, a user authenticates himself via a transaction-pseudonym to an organization (or another user) known as the verifier. The solutions focus on unilateral authentication. The basic functionality of the infrastructure is illustrated in Figure 6.2. Each of the two proposed schemes consists of five stages:

1. *Setup:* In the setup-stage the system parameters Γ are generated and each instance is initialized with them. The pseudonym certification authority generates a signature key-pair for the certification of pseudonyms. Furthermore, the revocation centers jointly generate shares of a secret key and the corresponding public key.

2. *User Registration:* In this stage, the user (or the pseudonym certification authority) generates the private user-identifier id and the corresponding public user-identifier ID. During a face-to-face authentication at the pseudonym certification authority, his personal data is stored (e.g. a copy of his passport) together with his public encryption key ek and the public user-identifier ID. In return he gets the system parameters Γ and the public keys of the pseudonym certification authority and of the set $\mathcal{R}$ of revocation centers.

3. *Establishment of Authentication Data:* In this stage, (parts of) the pseudonyms are derived from the public user-identifier. Furthermore they are signed by the pseudonym certification authority. Depending on the approach this is done within

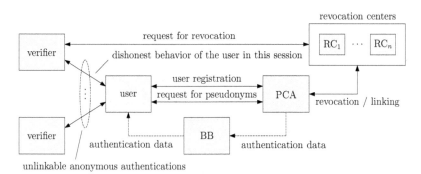

Fig. 6.2: Basic System Architecture.

a tamper-resistant device or using an appropriate blind signature scheme. A batch Λ of authentication data (signatures for l pseudonyms of the same user) is then encrypted using hybrid encryption with respect to the public encryption key of the user. The resulting encrypted batch Λ' is posted on the bulletin board or sent directly to the user.

4. *Authentication:* The verifier needs to have authentic access to Γ and the public keys of the pseudonym certification authority and $\mathcal{R}$. All the proposed authentication protocols work essentially as stated in Figure 6.3: the user computes a pseudonym η and extracts the corresponding signature σ from the batch Λ. Then he sends the pair (η, σ) to the verifier. If the sent data is authentic, the verification succeeds and the user has to run an interactive proof of ownership for the pseudonym. In the end the verifier accepts the authentication process if for an authentic η the proof of ownership passed verification. Otherwise, he rejects the authentication process.

5. *Optional Anonymity Revocation and Linking:* Depending on the information contained in the pseudonym, the set of revocation centers can jointly revoke the anonymity of the user by disclosing ID. Furthermore, some of the used pseudonyms provide optional linking, if id is known.

This chapter assumes that an appropriate security infrastructure exists, providing confidential, integrity-protected and authentic communication between the players, thus enabling a highly abstract description of the system's functionalities.

6.3.4 Two Authentication Schemes

For simplicity, the two proposed authentication schemes are denoted by AS1 and AS2. Protocols are denoted by Pj, where $j \in \{1, 1b, 2, 2b, 3\}$ if the corresponding authenti-

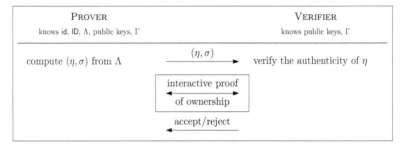

Fig. 6.3: Structure of an Authentication Protocol.

cation process is linkable over the private user-identifier id. If $j \in \{1a, 2a\}$ then the authentication process is not identifiable using id, or only if all pseudonyms of the system are disclosed. The latter we prohibit through the privacy policy.

The first authentication scheme, published in [SS06], is based on various problems related to the Discrete Logarithm Problem and the Double Discrete Logarithm Problem in groups of prime order. All the three protocols proposed there provide linkability if id is known. However, id can only be disclosed by $\mathcal{R}$ through the transcript of one correct execution of protocol AS1.P3. The anonymity can be revoked analogously through protocols AS1.P2 and AS1.P3. Depending on the view of the adversary, the security of AS1 can be reduced to the Discrete Logarithm Problem, the Diffie-Hellman Problem, the Decision Diffie-Hellman Problem or the Double Discrete Logarithm Problem in the groups $\mathbb{G}_q$ and $\mathbb{G}_p$, respectively.

Unfortunately, the design principles used for AS1 lead to the construction of rather inefficient instances of the protocols AS1.P2 and AS1.P3:

1. The given interactive proofs of ownership use a binary challenge space and hence require a large amount of exponentiations on the user-side.

2. Basing the protocols on an elliptic curve subgroup brings no advantage, because the order of the group must be large (approximately a length of 1024 bit for the current lower security bound) to render the Diffie-Hellman Problem intractable in the group from which the scalars are chosen.

As AS1 is not useful for smartcard-based applications, a further scheme is proposed in this chapter, AS2, which overcomes the drawbacks of the basic one in [SS06].

The key-idea of AS2 is to escrow id during the user's registration, so that protocol P3 is no longer necessary. Moreover, protocol P2 can be improved such that it is no longer based on the Double Discrete Logarithm Problem. Interactive proofs can be designed with a large challenge space so that the number of operations can be kept low by applying Damgård's parallelization techniques [Dam00]. As a by-product, the whole scheme can be based on an elliptic curve subgroup of prime order q. Generally, the new instances of protocol P1 and P2 can be realized in two ways:

1. Protocols A2.P1a and A2.P2a are designed, such that linking is infeasible.

2. Protocols A2.P1b and A2.P2b are designed, such that they are linkable using id.

Both schemes require the pseudonym certification authority to be based on a tamper-resistant device. In AS2, the user is provided with a smartcard so that a blind signature scheme could also be used instead. As a consequence, the pseudonym certification authority does not need to be based on a tamper-resistant device, but consequently establishing pseudonyms becomes quite cumbersome. To keep the discussion simple, further details on such a variant are not given in this work.

6.3.5 Security and Privacy

To prove the security and privacy of AS1 and AS2, it must be shown that requirements R1-R4 stated in Section 6.1.1 are fulfilled sufficiently. For R1-R3, the following attacks must not be efficiently mountable using a poly-bounded algorithm:

1. *Forging:* Generally, there are three ways to forge an authentication process:
   (a) Find a trapdoor in the pseudonym generation process.
   (b) Forge a signature of the pseudonym certification authority.
   (c) Forge the interactive proof of ownership used in the authentication process.

2. *Breaking the Anonymity:* Given the protocol transcript, obtain ID (or vice versa).

3. *Breaking the Unlinkability:* Given a set of protocol transcripts, identify the pseudonyms derived from the same user-identifier (or vice versa).

For AS1 and AS2, it is shown that the above stated attacks imply solving some well known problems that are believed to be hard. To fulfill requirement R4, the anonymity revocation process and linking must both be proven to be unique.

6.4 Authentication Scheme AS1

This section extends and proves the security of the approach given in [SS06].

6.4.1 Setup

A central initialization instance generates primes p and q, such that $p = 2q + 1$. Furthermore, the groups $\mathbb{G}_p$ and $\mathbb{G}_q < \mathbb{Z}_p^*$ are chosen with the orders p and q, respectively. The descriptions of the selected groups are stored in the system parameters Γ, including the generators g (of $\mathbb{G}_p$) and γ (of $\mathbb{G}_q$). Each instance of the system is initialized with Γ. The pseudonym certification authority generates a signature generation key sk_P in a tamper-resistant device and the corresponding verification key vk_P (it suffices to use an abstract view of the signature scheme). By using the techniques given in Section 5.2.6.1, the revocation centers jointly generate shares of an ElGamal decryption key $z_R \in \mathbb{Z}_q$ and reconstruct the corresponding encryption key $h_R = \gamma^{z_R}$. Hence, each RC_i holds a share $[z_{R_i}]^q$ of the decryption key and the set of commitments $\{h_{R_i} = \gamma^{[z_{R_i}]^q}\}_{i \in I(\mathcal{R})}$. The public keys are published on the bulletin board.

6.4.2 User Registration

The user performs a face-to-face authentication at the pseudonym certification authority, during which his personal identification data PDATA is captured (e.g. a copy of his passport). Then, he receives a smartcard initialized with Γ, vk_P, h_R and the Integrated ChipCard Serial Number (ICCSN). He then generates a key-pair (ek, dk) for asymmetric encryption (not specified in detail here) and the private user-identifier $\mathsf{id} \in \mathbb{Z}_p$ by using iCFNG. Finally, he sends (ek, dk) and the public user-identifier $\mathsf{ID} = g^{\mathsf{id}}$ to the pseudonym certification authority, which signs (ID, ek) resulting in the signature σ_{ID} and stores the vector $(\mathsf{ID}, ek, \sigma_{\mathsf{ID}}, \mathsf{PDATA})$ in a database.

6.4.3 Establishment of Authentication Data

The user contacts the pseudonym certification authority and performs an authentic request for a batch Λ, consisting of certified data for a correct execution of l authentication processes. By use of ID, the pseudonym certification authority retrieves $(ek, \sigma_{\mathsf{ID}})$ from the database and sends it to the tamper-resistant device. There, the signature is verified such that a corrupted pseudonym certification authority is identified. Then, l pairs of the form (b, σ) are generated as follows by the tamper-resistant device:

1. Compute $b = \mathsf{iCFNG}()$.

2. Compute $S((g^b, \mathsf{ID}^b), sk_P) = \sigma$.

The tamper-resistant device encrypts the batch Λ of l pairs of the form (b, σ) with ek, resulting in the ciphertext-batch Λ'. The pair (ID, Λ') is posted on the bulletin board.

6.4.4   Authentication Protocols

The basic structure of each authentication protocol is as follows (cf. Figure 6.4, top): at the beginning, the user needs to decrypt any pair (b, σ) from Λ'. Then he computes the pseudonym η. The pair (η, σ) is transferred to the verifier, who checks if η (or in some cases at least parts of it), σ and vk_P mutually correspond. If the verification succeeds, the verifier is convinced that the pseudonym has been issued by the pseudonym certification authority. The user must still prove ownership of η. The three provided variants are shown in Figure 6.4. Based on the verifications, the verifier informs the prover if he accepts or rejects the request.

6.4.4.1   Protocol AS1.P1

In this protocol the user only wants to show that he is a registered user, but as long as no information about id is disclosed, the authentication process can neither be linked to the user, nor any of his other authentication processes. Importantly, it needs to be mentioned that if id has been disclosed in any other authentication process of the same user, then even executions of AS1.P1 done by the same user can be identified. This comes from the fact, that the pseudonym $\eta = (\eta_1, \eta_2)$ is defined as follows:

$$\eta = (g^b, \mathsf{ID}^b)$$

Lemma 6.4.1 (η_1, η_2) *uniquely defines* ID.

Proof. It has to be shown that every Diffie-Hellman Triple in $\mathbb{G}_p$ is unique. Let $(A, B, C), (D, E, F) \in \mathbb{G}_p^3$ be two Diffie-Hellman Triples with respect to a generator g. If $(A, B) = (D, E)$, then $(dlog_g(A), dlog_g(B)) = (dlog_g(D), dlog_g(E))$ holds due to the injectivity of group exponentiation. But then $dlog_g(A)dlog_g(B) = dlog_g(D)dlog_g(E)$ must hold, since p is prime. Because b and id are elements of $\in \mathbb{Z}_p$, $(g^b, g^{\mathsf{id}}, g^{\mathsf{id}b})$ is a correct Diffie-Hellman Triple and (η_1, η_2) uniquely defines ID. $\square$

Obviously, if id is known, the equality check $\eta_1^{\mathsf{id}} = \eta_2$ succeeds and correspondence to the user can be identified without the explicit disclosure of the public user-identifier of every η. However, if id has not yet been disclosed in the system, there is no efficient chance better than solving the Discrete Logarithm Problem or the Divisible Diffie-Hellman Problem (both in $\mathbb{G}_p$) to identify the user. Apart from the structure of the pseudonym, this holds because a Σ-proof for the knowledge of a discrete logarithm is used. The user can either prove that he knows b, id or id$b$. The blinding b leaves the tamper-resistant device encrypted by ek. This guarantees that only the holder of dk can obtain b. In the current solution, the user proves that he knows b. Using Damgård's techniques [Dam00], the proof can be parallelized completely, such that only one exponentiation in $\mathbb{G}_p$ is necessary on user-side.

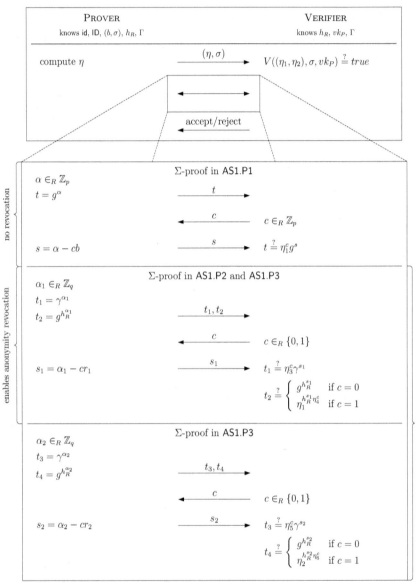

Fig. 6.4: Authentication Protocols AS1.P1, AS1.P2 and AS1.P3.

6.4.4.2 Protocol AS1.P2

If b is known in addition to η_2, then ID can be easily extracted by computing $\eta_2^{b^{-1}}$. To provide *optional* disclosure, b needs to be included in η in encrypted form. Since a user is generally not expected to be trustworthy, he has to prove that the corresponding ciphertext truly contains sufficient information to recover b. One solution to this problem is the use of verifiable encryption of the inverse of a discrete logarithm as given in Section 6.2.4. In this case, b^{-1} is ElGamal-encrypted with h_R and leads to the following definition of η:

$$\eta = (g^b, \mathsf{ID}^b, \gamma^{r_1}, b^{-1}h_R^{r_1}), \quad r_1 \in_R \mathbb{Z}_q$$

Corollary 6.4.1 $(\eta_1, \ldots, \eta_4)$ *uniquely defines* ID.

Proof. Directly follows from Lemma 6.4.1. □

The protocol given in Section 6.2.4 allows the user to show that (η_3, η_4) is an ElGamal ciphertext of the plaintext $dlog_g^{-1}(\eta_1)$. Again, parallelization can be achieved using Damgård's techniques. Unfortunately, only the number of sent message can be decreased, but not the number of exponentiations. This problem arises because of the binary challenge space (cf. Section 6.4.7).

6.4.4.3 Protocol AS1.P3

Decrypting b results in an efficient extraction of ID. However, b and ID are not sufficient to find any other pseudonym derived from ID. For an efficient linking, id is necessary. This can be achieved by additionally including the ElGamal-encryption of $(\mathsf{id}b)^{-1}$:

$$\eta = (g^b, \mathsf{ID}^b, \gamma^{r_1}, b^{-1}h_R^{r_1}, \gamma^{r_2}, (\mathsf{bid})^{-1}h_R^{r_2})$$

where $r_1, r_2 \in_R \mathbb{Z}_q$. Hence, in addition to the second protocol, one more run of the protocol given in Section 6.2.4 is necessary. If anonymity revocation and linking are always both necessary, then it suffices to contain id^{-1} in η in encrypted form, so that only one execution of the Σ-proof is necessary.

Corollary 6.4.2 $(\eta_1, \ldots, \eta_6)$ *uniquely defines* ID.

Proof. Directly follows from Lemma 6.4.1. □

The verifiable encryption of $(\mathsf{id}b)^{-1}$ can be removed if id is escrowed (in encrypted or shared form) at the revocation centers. This improvement causes some changes in the linking process. A solution is given in the description of AS2.

6.4.5   Revocation

When a user behaves dishonest after an accepted run of protocol AS1.P2 or AS1.P3, the anonymity can be revoked by protocol AS1.DisclosePublicIdentifier (cf. Figure 6.5). Therefore, (η_3, η_4) is sent to the revocation centers, who execute the ElGamal threshold decryption protocol (cf. Section 5.2.6.2), to obtain b^{-1}. ID can be extracted from η_2 and used to identify the user in cooperation with the pseudonym certification authority.

Protocol AS1.DisclosePublicIdentifier

Input (RC$_i$): $[z_{R_i}]^q$, (η_3, η_4)

Output (PCA): ID, b^{-1}, PDATA

1. The revocation centers participate in a run of protocol ElGamal-TDEC to decrypt (η_3, η_4), resulting in the plaintext b^{-1}.

2. Finally, ID $= \eta_2^{b^{-1}}$ is sent to the PCA, who can identify the corresponding user.

Protocol AS1.LinkPseudonyms

Input (RC$_i$): $[z_{R_i}]^q$, (η_5, η_6), b^{-1}

Output (RC$_i$): N_{ID}

1. In addition to running AS1.DisclosePublicIdentifier, the revocation centers participate in a run of ElGamal-TDEC to decrypt (η_5, η_6), and get the plaintext $(\mathsf{id}b)^{-1}$.

2. Then, id can be computed as follows: id $= b^{-1}((\mathsf{id}b)^{-1})^{-1}$.

3. Each revocation center forms the set N_{ID} of pseudonyms which correspond to ID:

$$N_{\mathsf{ID}} = \{\eta \mid \eta_1^{\mathsf{id}} = \eta_2 \ \wedge \ \eta \in N \ \wedge \ \mathsf{id} = dlog_g(\mathsf{ID})\}$$

Fig. 6.5: Anonymity Revocation and Linking.

Given the protocol transcript of a run of AS1.P3 and ID, the revocation centers can jointly compute id and then identify all derived pseudonyms in the set N of all pseudonyms. Therefore, they execute protocol AS1.LinkPseudonyms (cf. Figure 6.5).

A problem which has not been discussed in the preliminary version [SS06] is the problem of how to pool pseudonyms. Section 6.6 is dedicated to this topic.

6.4.6   Security and Privacy

This section provides proofs that under several computational assumptions, AS1 fulfills requirements R1-R4 as stated in Section 6.1. Therefore, it needs to be shown that the attacks stated in Section 6.3.5 are infeasible for a poly-bounded attacker. Notice that the following always includes the strongest assumption in theorems, since it covers weaker ones as a by-product. For instance, assuming that the Diffie-Hellman Problem is hard implies that the Discrete Logarithm Problem is hard too.

Lemma 6.4.2 *Given η_1, η_2, g and p, computing ID, such that $\eta_2 = \mathsf{ID}^{\,d\log_g(\eta_1)}$ holds, is equivalent to solving the Divisible Diffie-Hellman Problem in $\mathbb{G}_p$.*

Proof. Given g^x, g^y and g, the goal is to compute $g^{xy^{-1}}$. By setting $x = \mathsf{id}b$ and $y = b$, $xy^{-1} = \mathsf{id}bb^{-1} = \mathsf{id}$. $\quad\square$

Theorem 6.4.1 *With the assumption that solving the Diffie-Hellman Problem in $\mathbb{G}_q$ and $\mathbb{G}_p$ is hard, authentication scheme AS1 fulfills R1 if at most $t < n/2$ revocation centers are actively dishonest.*

Proof. It must be shown that the attacks (a)-(c) of Section 6.3.5 are infeasible:

(a) For protocol AS1.P1, given $\eta = (g^b, \mathsf{ID}^b)$, the goal is to find ID, which is equivalent to solving the Diffie-Hellman Problem in $\mathbb{G}_p$. In protocol AS1.P2, the adversary additionally knows $(\eta_3, \eta_4) = (\gamma^{r_1}, b^{-1}h_R^{r_1})$ and so the goal is to find b or r_1. Both require solving the Diffie-Hellman Problem in $\mathbb{G}_q$. Another strategy can be to compute z_R from h_R which is equivalent to the Discrete Logarithm Problem in $\mathbb{G}_q$, or compromise at least $t + 1$ revocation centers to reconstruct z_R and thus obtain b^{-1}. The security analysis of protocol AS1.P3 can be sketched analogously.

(b) Depends on the signature scheme used.

(c) The strategy of an adversary is to use a copy of a valid transcript of the applicable Σ-proof to authenticate to a verifier on behalf of the original user. The Σ-proofs [Sch89, Sta96] used here withstand such an attack, since they are zero-knowledge with respect to a dishonest verifier if Damgård's techniques are applied. $\quad\square$

Lemma 6.4.3 *Given η_1, η_2, g and p, computing id, such that $\eta_2 = \eta_1^{\mathsf{id}}$ holds, is equivalent to solving the Discrete Logarithm Problem in $\mathbb{G}_p$.*

Proof. Trivial. $\quad\square$

Theorem 6.4.2 *With the assumption that solving the Diffie-Hellman Problem and the Double Discrete Logarithm Problem with respect to $\mathbb{G}_q$ and $\mathbb{G}_p$ are hard, authentication scheme AS1 fulfills R2 if at most $t < n/2$ revocation centers are actively dishonest.*

Proof. Assuming that the correspondence between ID and the user is publicly known, the goal of an attack to break the anonymity is to link ID to an authentication process. During the establishment of the authentication data, this requires breaking the tamper-resistant device or breaking the hybrid cryptosystem by which batch Λ is encrypted. At the beginning of every authentication protocol, the pseudonym and the corresponding signature are sent to the verifier. Anonymity is preserved for all three protocols:

P1: Given $\eta = (g^b, \mathsf{ID}^b)$ and the protocol transcript, the goal is to find ID. From Lemma 6.4.2 and [Sch89], the problem can be reduced to the Divisible Diffie-Hellman Problem and the Discrete Logarithm Problem in $\mathbb{G}_p$, respectively.

P2: The ciphertext $(\gamma^{r_1}, b^{-1}h_R^{r_1})$ is also provided and so the problem of finding ID is equivalent to the Divisible Diffie-Hellman Problem in $\mathbb{G}_p$ (Lemma 6.4.2), the Diffie-Hellman Problem in $\mathbb{G}_q$ (ElGamal), the problem of breaking the verifiable encryption scheme stated in Section 6.2.4 (i.e. Double Discrete Logarithm Problem) and the problem of compromising at least $t + 1$ revocation centers.

P3: The security analysis works analogous to that of protocol AS1.P2.   □

So far, it has been shown that the identity of the user cannot be efficiently extracted from a pair (g^b, ID^b) if the Discrete Logarithm Problem and the Divisible Diffie-Hellman Problem are hard in $\mathbb{G}_p$. However, it has been assumed that only the considered pseudonym and the system parameters are accessible. It also needs to be shown, that even if all public user-identifiers are available to the adversary, pseudonyms can neither be linked to their holder nor mutually to each other.

Lemma 6.4.4 *Let I be the set of all public user-identifiers and N the set of pairs of the form (g^b, ID^b), where $b \in \mathbb{Z}_p$ and $\mathsf{ID} \in I$. Given any $(\eta_1, \eta_2) \in N$, deciding to which element of I it corresponds is computationally equivalent to solving the Decision Diffie-Hellman Problem in $\mathbb{G}_p$.*

Proof. First, assume that an oracle $\mathcal{O}_{\mathrm{DDP}}$ exists, which on the input of a triple $(X, Y, Z) \in \mathbb{G}_p^3$, returns 1 if it is a Diffie-Hellman Triple with respect to g or 0 otherwise. Querying $\mathcal{O}_{\mathrm{DDP}}((\mathsf{ID}, \eta_1, \eta_2), g)$ for each $\mathsf{ID} \in I$ gives 1 exactly once. This directly follows from Lemma 6.4.1. For the converse let $\mathcal{O}$ be an oracle which on input I, (η_1, η_2) and g returns $\mathsf{ID} \in I$ such that $\eta_2 = \mathsf{ID}^{dlog_g(\eta_1)}$ if one exists or $\bot$ otherwise. Obviously, if an element ID has been returned then $(\mathsf{ID}, \eta_1, \eta_2)$ must form a Diffie-Hellman Triple, since $\mathsf{ID} = g^{\mathsf{id}}$ and $\eta = (g^b, \mathsf{ID}^b)$. Hence, given any triple $(X, Y, Z) \in \mathbb{G}_p^3$, the Decision Diffie-Hellman Problem can be solved by computing $\mathcal{O}(\{X\}, (Y, Z), g) = X'$, where $X' = X$ iff (X, Y, Z) forms a Diffie-Hellman Triple and $X' = \bot$ otherwise.   □

Finally, it needs to be shown that given the set of all pseudonyms of the system, a poly-bounded algorithm cannot decide which of them belong to the same user-identifier.

Lemma 6.4.5 *Let I and N be as given in Lemma 6.4.4. Determining any set $N' \subseteq N$ whose elements are derived from the same user-identifier is at least as hard as solving the Decision Diffie-Hellman Problem in $\mathbb{G}_p$.*

Proof. It has to be shown, that deciding whether two given elements of N are derived from the same public user-identifier implies solving the Decision Diffie-Hellman Problem in $\mathbb{G}_p$. Let $\mathcal{O}$ be an oracle which on input of $(g^{b_i}, \mathsf{ID}^{b_i}), (g^{b_j}, \widehat{\mathsf{ID}}^{b_j}) \in N$ and g outputs 1 iff $\mathsf{ID} = \widehat{\mathsf{ID}}$ or 0 otherwise. Now it must be shown that $\mathcal{O}$ can be used to decide whether $(X, Y, Z) \in \mathbb{G}_p^3$ forms a Diffie-Hellman Triple or not. Querying $\mathcal{O}((g, X), (Y, Z), g)$ results in the output 1 iff $z = xy$ and thus (X, Y, Z) a Diffie-Hellman Triple or 0 otherwise. $\square$

Notice that if I is also given, one can shown that the above problem is even computationally *equivalent* to the Decision Diffie-Hellman Problem. Therefore, one has just to apply Lemma 6.4.4 for all combinations of I and N.

Theorem 6.4.3 *With the assumption that the Decision Diffie-Hellman Problem in $\mathbb{G}_p$ and the Diffie-Hellman Problem in $\mathbb{G}_q$ are hard, authentication scheme* AS1 *fulfills R3 if at most $t < n/2$ revocation centers are actively dishonest.*

Proof. Given a set of public user-identifiers and a set of pairs of the form (g^b, ID^b), linking is not feasible due to Lemmata 6.4.4 and 6.4.5. The linking between a user request and generated pseudonyms or signatures is hidden by the use of the tamper-resistant device and the fact that its encrypted output can only be decrypted by the associated user. For efficient linking, id is necessary which cannot be determined from the public one (cf. Lemma 6.4.3). Concerning also the authentication protocols, only the third can lead to a disclosure of id. This implies solving the Diffie-Hellman Problem in $\mathbb{G}_q$ (ElGamal) or compromising at least $t + 1$ revocation centers. $\square$

Theorem 6.4.4 *Authentication scheme* AS1 *fulfills R4 if at most $t < n/2$ revocation centers are actively dishonest.*

Proof. From Corollary 6.4.1 anonymity revocation leads to the unique identification of a user if at least $t + 1$ revocation centers are honest. For a successful linking, the correct id must be extractable from η used for AS1.P3. This works correctly if at least $t + 1$ revocation centers are honest. Given id, every corresponding pseudonym can be uniquely identified (follows from Lemma 6.4.1 and Corollaries 6.4.1 and 6.4.2). $\square$

6.4.7 Efficiency

Communication: Only one user request is necessary for each batch of pseudonyms. The authentication process is a three-way protocol if Damgård's techniques [Dam00] are used to parallelize the Σ-protocols.

Time: The generation of the linking information requires two exponentiations in $\mathbb{G}_p$. The first proof of ownership is based on a standard Σ-proof for the knowledge of a discrete logarithm which using parallelization can be reduced to one exponentiation on the user-side while remaining zero-knowledge even in the presence of a dishonest verifier. The second protocol requires two exponentiations in $\mathbb{G}_q$ for the generation of the ciphertext. Although the proof of knowledge used can be parallelized, the number of exponentiations cannot be reduced because a binary challenge space is used. Hence, for 80-bit security with respect to a cheating prover, approximately 160 exponentiations in $\mathbb{G}_q$ and 80 exponentiations in $\mathbb{G}_p$ are performed on the user-side. The third protocol has a similar problem since the same type of Σ-proof is used (in the basic variant twice). Hence, approximately 320 exponentiations in $\mathbb{G}_q$ and 160 exponentiations in $\mathbb{G}_p$ are performed on the user-side. Obviously, these two protocols are not very practical (especially when smartcards are involved).

Space: With respect to the recommendations given in Table 3.4, each pseudonym requires storing the blinding (1024 bits) and a signature (320 bits if the Digital Signature Standard is used). Since the proofs of knowledge are run in parallel, the first-messages of all rounds have to be generated. For the second protocol, this requires approximately 160KB (for the 160 exponentiations in $\mathbb{G}_q$) of temporary space.

6.5   Authentication Scheme AS2

The main-goal of this section is to present the design of an authentication scheme which has similar properties as AS1, but is optimized such that the user-side of each authentication protocol can be run on a power-limited device (e.g. a smartcard). Thus, it is an intrinsic requirement to keep the complexity of communication, time and space low. Additionally, some modifications take place regarding the privacy of the user. The following sketches a short overview of the improvements prior to giving more details.

Efficiency (Communication): As in AS1, Σ-proofs are parallelized.

Efficiency (Time): Anonymity revocation in AS1 requires the disclosure of ID. Hence, it suffices to include an encryption of ID instead of b^{-1} in the pseudonym at the beginning of protocol P2. The Σ-proof of the basic scheme needs to be replaced by proofs for the knowledge and equality of discrete logarithms. Such proofs, if parallelized, require only a few exponentiations on the user-side because the challenge space is exponential. A further speed-up can be achieved by basing the whole scheme on an elliptic curve group $\mathbb{G}_q$ of small prime order q. This is possible because none of the protocols are based on the Double Discrete Logarithm Problem.

Efficiency (Space): Basing the scheme on an elliptic curve group also leads to minimal memory requirements. As one can see later on, a run of protocol AS2.P2 requires less than 1 KB of temporary space. Furthermore, the following optimization can be made to the space of a batch of l pseudonyms: if the ideal collision-free number generator used is a pseudo-random number generator, then each blinding can be derived from the previous state(s). Furthermore, if the Digital Signature Standard is used, for instance, every signature requires only about 320 bits of space.

Perfect Unlinkability for Selected Pseudonyms: Once id has been disclosed in the basic scheme, all corresponding pseudonyms are identifiable. In AS2, the linkability of a pseudonym can be activated by the user himself at the beginning of the authentication process. If not activated for protocol P1, linking is not possible even if the attacker is unbounded. If P2 is used, linking is theoretically possible, but requires the revocation of *all* the pseudonyms issued. In practice, the majority of the revocation centers will not agree to do so, since this would violate the privacy policy.

Selective Shared Linking of Pseudonyms: Protocol AS1.P3 has been used to additionally provide linking. This protocol can be completely removed by a slight modification of the scheme: the user needs to share id $\in \mathbb{Z}_q$ over $\mathcal{R}$ by using Feldman's verifiable secret sharing (cf. Section 5.2.4.2). In the end of this escrow process, every RC_i holds a poly-share $[\mathsf{id}_i]^q$ of id and the set of commitments $\{\mathsf{ID}, \mathsf{ID}_i = g^{[\mathsf{id}_i]^q}\}_{i \in I(\mathcal{R})}$ which lie in an appropriate group $\mathbb{G}_q$ of prime order q (in the following an elliptic curve group). For an optional linking, the anonymity has to be revoked first resulting in ID, which each RC_i can use to identify the share $[\mathsf{id}_i]^q$. Afterwards, shared linking can be performed with a selected set of pooled pseudonyms (cf. Section 6.6). Since a qualified subset of $\mathcal{R}$ is necessary, the selected pooled pseudonyms (on which the revocation centers agreed to run the linking process) are analyzed exclusively. Such *controlled* linking is not possible in the basic scheme, even if threshold decryption is used: once id has been disclosed, *every* revocation center can perform linking-verifications on *any* pseudonym. Since up to t (actively) dishonest revocation centers are tolerated, unauthorized and uncontrolled linking is possible in AS1 after the disclosure of id. This is not so in the extended scheme because id is never reconstructed.

6.5.1 Setup

Let $E(\mathbb{Z}_p)$ be an elliptic curve group, where p is an odd prime. Furthermore, let $\mathbb{G}_q < E(\mathbb{Z}_p)$ with prime order q, such that $q|\#E(\mathbb{Z}_p)$. A point P is chosen such that $\langle P \rangle = \mathbb{G}_q$. Each instance of the system is initialized with the system parameters contained in Γ, including the domain parameters of $\mathbb{G}_q$.

Using the techniques of Section 5.2.6.1, the revocation centers jointly generate shares of the ElGamal decryption key $z_R \in \mathbb{Z}_q$. Each player is provided with a share of z_R, the encryption key $H_R = Pz_R$ and the set of commitments $\{H_{R_i} = P[z_{R_i}]^q\}_{i \in I(\mathcal{R})}$. As usual, the tamper-resistant device of the pseudonym certification authority generates (sk_P, vk_P). The public keys are posted on the bulletin board.

6.5.2 User Registration

The user authenticates face-to-face himself with the pseudonym certification authority and receives Γ. He then uniquely generates $\mathsf{id} \in \mathbb{Z}_q$ and shares it over the set $\mathcal{R}$ of revocation centers using Feldman's verifiable secret sharing (here based on $\mathbb{G}_q$). Hence, each revocation center RC_i is provided with a share $[\mathsf{id}_i]^q$ and the set of commitments $\{\mathsf{ID}, \mathsf{ID}_i = P[\mathsf{id}_i]^q\}_{i \in I(\mathcal{R})}$. This data is stored locally by each RC_i. It is assumed that the pseudonym certification authority is connected to the broadcast channel used for the escrow process of id and hence receives ID as well. The user generates the pair (ek, dk) and sends ek to the pseudonym certification authority. The tamper-resistant device signs ID and ek as done in $\mathsf{AS1}$ and stores $(\mathsf{ID}, ek, \sigma_{\mathsf{ID}}, \mathrm{PDATA})$ in the database.

6.5.3 Establishment of Authentication Data

The pseudonym certification authority certifies one part of a generated ElGamal ciphertext. Thus, the tamper-resistant device generates randomizers for ElGamal encryptions instead of the blindings. For the sake of efficiency, the randomizers are generated by a pseudo-random collision-free number generator. This has the advantage, that the user only needs its initial state to be able to derive all the randomizers of the current batch. At first, the tamper-resistant device initializes iCFNG with a random seed s. Then, l signatures σ are generated as follows:

1. Compute $r = \mathsf{iCFNG}()$.

2. Compute $S(\mathsf{ID} + H_R r, sk_P) = \sigma$

Then the batch Λ, containing l signatures and the random seed s, is encrypted and published analogously to $\mathsf{AS1}$ through the pair (ID, Λ').

Remark 6.5.1 If a blind signature scheme is used, step 1 is done in the smartcard of the user. The used collision-free number generator can hence be based on the Integrated ChipCard Serial Number. Step 2 is done according to the blind signature scheme used. Here, the pseudonym certification authority must be convinced that it communicates with the smartcard. This can be achieved by providing each smartcard with an authentic signature key-pair.

6.5.4 Authentication Protocols

Since id has been escrowed during the registration process, only two authentication protocols are necessary. In their basic variants, both protocols (denoted by $\mathsf{P1a}$ and $\mathsf{P2a}$) do not provide linkability. However, if necessary, the user can verifiably activate

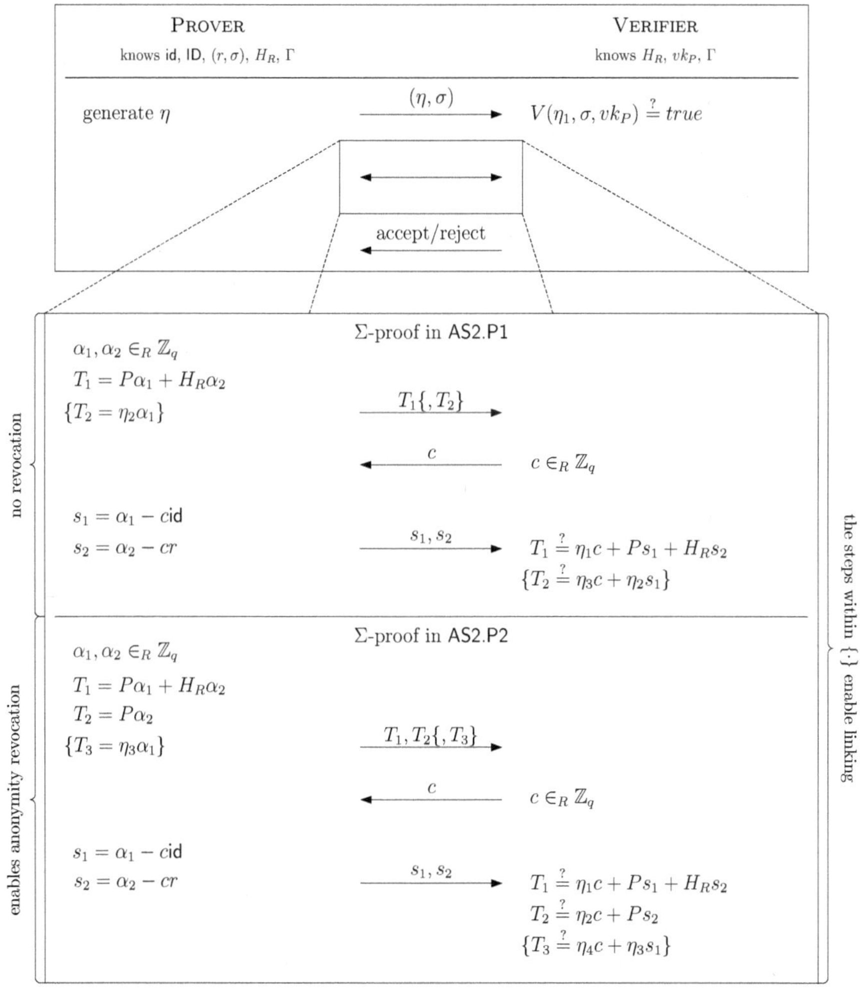

Fig. 6.6: Authentication Protocols AS2.P1 and AS2.P2.

the linking property of the authentication protocol (the corresponding protocols are denoted by P1b and P2b). Activating linkability leads to some additional components contained in the pseudonyms used and some additional steps for the protocols. These components and steps are placed within curly brackets $\{\cdot\}$.

6.5.4.1   Protocol AS2.P1

Prior to generating a pseudonym, iCFNG has to be initialized properly such that the valid sequence (the same as computed by the pseudonym certification authority) of randomizers can be generated efficiently. The pseudonym η is hence defined as follows:

$$\eta = (\text{ID} + H_R r\{, Pb, \text{ID}b\}), \quad r = \text{iCFNG}(), \quad b \in_R \mathbb{Z}_q$$

Lemma 6.5.1 η_1 defines ID with computational uniqueness.

Proof. If the authentication protocol is based on η_1, only the user can disclose his own identity (e.g. in applications where the user has a strong interest in doing so). Disclosing an identity $\widehat{\text{ID}} \neq \text{ID}$ requires the knowledge of z_R, which requires solving the Discrete Logarithm Problem in $\mathbb{G}_p$ or compromising $t+1$ revocation centers. $\square$

Corollary 6.5.1 $(\eta_1, \ldots, \eta_3)$ uniquely defines ID.

Proof. This directly follows from Lemma 6.4.1 applied to $\mathbb{G}_q$. $\square$

Protocol AS2.P1a is based on η_1 exclusively and hence requires two parallel runs of a Σ-proof for the knowledge of a discrete logarithm based on the same challenge. Protocol AS2.P1b is based on (η_1, η_2, η_3) and thus additionally requires one proof of the equality of discrete logarithms. Using Damgård's techniques, the proofs can be parallelized and hence only require a few scalar multiplications on the user-side.

6.5.4.2   Protocol AS2.P2

If Pr is known in addition to $\text{ID} + H_R r$, the revocation centers can jointly compute ID using ElGamal threshold decryption. Hence, η is defined as follows:

$$\eta = (\text{ID} + H_R r, Pr\{, Pb, \text{ID}b\}), \quad r = \text{iCFNG}(), \quad b \in_R \mathbb{Z}_q$$

Lemma 6.5.2 (η_1, η_2) uniquely defines ID.

Proof. Assume that a public user-identifier $\widehat{\text{ID}} \neq \text{ID}$ exists, such that $(\eta_1, \eta_2) = (\widehat{\text{ID}} + H_R r, Pr)$ holds. Then, $\eta_1 = P(\text{id} + z_R r) = P(\widehat{\text{id}} + z_R r)$ and thus $\text{id} + z_R r = \widehat{\text{id}} + z_R r$, which in turn implies that $\text{id} = \widehat{\text{id}}$ holds. This, however, yields a contradiction. $\square$

Corollary 6.5.2 $(\eta_1, \ldots, \eta_4)$ uniquely defines ID.

Proof. This directly follows from Lemma 6.5.2. $\square$

The pseudonym $\eta = (\eta_1, \eta_2)$ is an ElGamal ciphertext which, if jointly decrypted by the revocation centers, results in ID. Obviously, given a set of pseudonyms of this form, linking is possible if all pairs are decrypted. If $(\eta_1, \ldots, \eta_4)$ is used, then linking is selectively possible for a given private user-identifier id. The Σ-proofs for AS2.P2 can be designed analogously to the ones of protocol AS2.P1.

6.5.5 Revocation

If protocol AS2.P2 is used, anonymity can be revoked by simply decrypting (η_1, η_2) using ElGamal threshold decryption (cf. protocol AS2.DisclosePublicIdentifier, Figure 6.7). Let $j = |\eta| - 1$. Then, given a particular public user-identifier ID and a set N of pairs of the form (η_j, η_{j+1}), the linkable pseudonyms derived from ID can be identified (cf. protocol AS2.LinkPseudonyms). Therefore, the revocation centers have to jointly check if $\eta_{j+1} = \eta_j$id holds without reconstructing id. Hence, each RC_i has to broadcast $\eta_{j_i} = \eta_j[\text{id}_i]^q$ together with a proof that $ecdlog_P(\text{ID}_i) = ecdlog_{\eta_j}(\eta_{j_i})$ (note that every member of $\mathcal{R}$ has authentic access to every ID_i). If at least $t + 1$ correct proofs have been broadcast, the remainder of the verification can be done locally. The set of pseudonyms derived from ID can then be formed based on these results.

Protocol AS2.DisclosePublicIdentifier

Input (RC_i): $[z_{R_i}]^q$, (η_1, η_2)

Output (PCA): ID, PDATA

1. The revocation centers participate in a run of protocol ElGamal-TDEC (with respect to elliptic curves) to decrypt (η_1, η_2) resulting in the plaintext ID.
2. Having ID, the pseudonym certification authority can identify the user.

Protocol AS2.LinkPseudonyms

Input (RC_i): $[\text{id}_i]^q$, η, $j := |\eta| - 1$

Output (RC_i): N_{ID}

1. Each RC_i broadcasts $\eta_{j_i} = \eta_j[\text{id}_i]^q$ and a proof τ_i that $ecdlog_P(\text{ID}_i) = ecdlog_{\eta_j}(\eta_{j_i})$.
2. $\mathcal{Q}$ is defined based on the validity of the non-interactive proofs that were broadcast.
3. Each member of $\mathcal{Q}$ forms the set N_{ID} of pseudonyms corresponding to ID:

$$N_{\text{ID}} = \{\eta \mid \eta_{j+1} = \sum_{i \in I(\mathcal{Q})} \eta_{j_i}\lambda_i\}$$

Fig. 6.7: Anonymity Revocation and Linking.

6.5.6 Security and Privacy

Lemma 6.5.3 *The Σ-protocols in AS2.P1 and AS2.P2 satisfy (1) completeness, (2) special soundness, and (3) are special honest-verifier zero-knowledge.*

Proof. Without loosing generality, the completeness for AS2.P2b is considered. The properties (2) and (3) can be shown analogously to the proofs given in Section 5.2.3.

(1) If the prover and the verifier follow the specification, then the verification succeeds:

$$\eta_1 c + P s_1 + H_R s_2 = (\mathsf{ID} + H_R r)c + P(\alpha_1 - c\mathsf{id}) + H_R(\alpha_2 - cr)$$
$$= P\mathsf{id}c + H_R rc + P\alpha_1 + P(-c\mathsf{id}) + H_R \alpha_2 + H_R(-cr)$$
$$= P\alpha_1 + H_R \alpha_2 = T_1$$
$$\eta_2 c + P s_2 = P rc + P(\alpha_2 - cr)$$
$$= P rc + P\alpha_2 + P(-cr) = P\alpha_2 = T_2$$
$$\eta_4 c + \eta_3 s_1 = (\mathsf{ID}b)c + (Pb)(\alpha_1 - c\mathsf{id})$$
$$= P\mathsf{id}bc + Pb\alpha_1 + P(-bc\mathsf{id})$$
$$= Pb\alpha_1 = \eta_3 \alpha_1 = T_3 \qquad \qquad \square$$

Theorem 6.5.1 *With the assumption that solving the Diffie-Hellman Problem is hard in $\mathbb{G}_q$, authentication scheme AS2 fulfills R1 if at most $t < n/2$ revocation centers are actively dishonest.*

Proof. It has to be shown that the attacks (a)-(c) stated in Section 6.3.5 are infeasible:

(a) For protocol AS2.P1, given $\eta = (\mathsf{ID} + H_R r)$ and H_R, the goal is to find ID. Since η is a P-commitment this is impossible. If $\eta = (\mathsf{ID} + H_R r, Pb, \mathsf{ID}b)$, then ID can be obtained by solving the Divisible Diffie-Hellman Problem. As far as protocol AS2.P2 is concerned, the goal is to determine ID from $(\mathsf{ID}+H_R r, Pr)$ and H_R which is equivalent to solving the Diffie-Hellman Problem (ElGamal). Since the Diffie-Hellman Problem and the Divisible Diffie-Hellman Problem are computationally equivalent (cf. Section 4.2.3), it is not easier to forge AS2.P2 if a pseudonym of the form $\eta = (Pr, \mathsf{ID} + H_R r, Pb, \mathsf{ID}b)$ is used. Notice that forging is possible if at least $t + 1$ revocation centers have been compromised.

(b) Depends on the signature scheme used.

(c) From Lemma 6.5.3 it follows that the Σ-proofs are honest-verifier zero-knowledge. With Damgård's technique the zero-knowledge property can be achieved for dishonest verifiers. Hence, the protocol transcripts cannot be used to authenticate on behalf of the original user. If the prover is dishonest, the verifier accepts a Σ-proof with negligible probability, since the challenge space is exponential. $\qquad \square$

Theorem 6.5.2 *With the assumption that solving the Diffie-Hellman Problem in $\mathbb{G}_q$ is hard, authentication scheme AS2 fulfills R2 if at most $t < n/2$ revocation centers are actively dishonest.*

Proof. The establishment of authentication data can be conducted analogously to Theorem 6.4.2. It remains to show that a poly-bounded adversary cannot efficiently determine ID from the protocol information. For protocol AS2.P1 this means extracting

ID out of $\eta = (\text{ID} + H_R r)$ or out of $\eta = (\text{ID} + H_R r, Pb, \text{ID}b)$ and the corresponding protocol transcript. The first variant reveals no information about ID since it contains a P-commitment (cf. Theorem 6.5.1). From Lemma 6.5.3, the proof of ownership used is ensured to be zero-knowledge. The second variant provides anonymity with respect to the Divisible Diffie-Hellman Problem in $\mathbb{G}_q$ (computing ID from (η_3, η_4)). Protocol AS2.P2 for pseudonyms of the form $\eta = (\text{ID} + H_R r, Pr)$ provides computational hiding since η is an ElGamal ciphertext. If Pb and $\text{ID}b$ are also given, ID remains hidden due to the Divisible Diffie-Hellman Problem. □

Theorem 6.5.3 *With the assumption that solving the Decision Diffie-Hellman Problem in $\mathbb{G}_q$ is hard, authentication scheme AS2 fulfills R3 if at most $t < n/2$ revocation centers are actively dishonest.*

Proof. Linking in general requires the knowledge of ID. Then, each RC_i can identify the share $[\text{id}_i]^q$ and so id can be *used* in shared form if at least $t+1$ members of $\mathcal{R}$ agree. Pseudonyms of the form $\text{ID} + H_R r$ are not identifiable, since $H_R r$ is a random blinding. If Pr is also given, then z_R can be used to decrypt ID. To find every pseudonym of this form derived from ID, the revocation centers have to decrypt all existing pseudonyms. Per assumption (privacy policy) the revocation centers will not agree to do so, because this requires the decryption of *all* analyzed pseudonyms which violates the privacy of the system. If any of the pseudonyms contains $(Pb, \text{ID}b)$, then shared linking can be applied. This process leads to a unique identification of the pseudonyms without violating the privacy of the *other* users (cf. Section 6.5.5). If id is not given, then from Lemmata 6.4.4 and 6.4.5 applied to $\mathbb{G}_q$ and due to the indistinguishability of ElGamal ciphertexts, linking is infeasible due to the Decision Diffie-Hellman Problem. □

Theorem 6.5.4 *Authentication scheme AS2 fulfills R4 if at most $t < n/2$ revocations centers are actively dishonest.*

Proof. Decrypting (η_1, η_2) from a transcript of protocol AS2.P2 directly discloses ID uniquely (cf. Lemma 6.5.2). If the pseudonym is linkable, then it contains $(Pb, \text{ID}b)$ and thus can be uniquely identified using id in a distributed way. This works correctly if at least $t + 1$ revocation centers collaborate (cf. Figure 6.7). □

6.5.7 Efficiency

Communication: Analogous to AS1.

Time: All the protocols can be fully parallelized using Damgård's techniques in terms of communication *and* computations. The latter holds since a challenge space with exponential size can be used. Thus, at most 8 scalar multiplications (4 for the generation of η and 4 for the first-message of the Σ-proof) are necessary on user-side for the authentication process. This is a satisfying result for situations that require the use of smartcards. The computational costs for the linking process only increase with linear time based on the number of verified pseudonyms.

Space: If the Digital Signature Standard is is used for signing $ID + H_R r$, then a batch of l pseudonyms consists of 160 bits for the seed and approximately $320l$ bits for the signatures. A smartcard with a 16KB EEPROM could then store approximately 400 signatures. This enables the use of AS2 for smartcard-based applications.

6.6 Controlled Pooling of Pseudonyms

The linking techniques of AS1 and AS2 require that the (linkable) pseudonyms are pooled somewhere. Such a pooling also has the advantage that the double-use of a pseudonym can be detected which is very useful if it is used in electronic payment systems: if a virtual banknote is used more than once, the anonymity of the corresponding holder can be revoked. For pooling pseudonyms, we consider two variants:

1. *Pre-pooling:* Pooling of pseudonyms right after their creation.

2. *Post-pooling:* Pooling of pseudonyms that have already been used.

Pre-pooling requires a technique to ensure that there is no detectable link between a user request and a pseudonym. Post-pooling requires that after an authentication process, the verifier sends the received pseudonym to some kind of linking directory.

6.6.1 Pre-Pooling of Pseudonyms

Figure 6.8 illustrates the basic idea. The (certified) linking information of every pseudonym established needs to be encrypted with a public key, whose corresponding secret key is shared among the set of mixers. The resulting ciphertext is then stored locally in the vector ρ'. If sufficiently many encrypted pseudonyms of *different* user requests are collected, then ρ' is re-randomized using the mix-net and finally jointly decrypted. The resulting pool ρ contains all linking information in plain form.

Pooling in AS1: The tamper-resistant device performs the desired encryptions. Each pair (g^b, ID^b) is encrypted and then placed in ρ'. The remainder works as sketched above. Notice that linking is only possible if a private user-identifier has been disclosed.

Pooling in AS2: In this scheme, pooling of $ID + H_R r$ is useless, since for a correct linking (Pb, IDb) is necessary which, however, is generated at random by the user himself. Thus, in the current version of AS2, only pooling of (linkable) pseudonyms that have already been used makes sense (cf. next section).

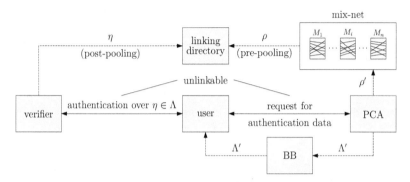

Fig. 6.8: Pooling of Pseudonyms.

Another solution can be to merge the ideas behind AS1 and AS2. The authentication data is generated and certified as described in AS1 but based on the system parameters of AS2. Then, the data is pooled as discussed above using mixers. Contrary to AS1, id is escrowed as described for AS2 so that the efficient authentication protocols AS2.P1b and AS2.P2b can be used. Thus, if initial pooling is necessary such that linking is always possible by $\mathcal{R}$, then this combined variant should be used.

A further advantage of initial pooling is that if a user turned out to be dishonest, all his pseudonyms can be identified and invalidated.

6.6.2 Post-Pooling of Pseudonyms

This variant of pooling is necessary if linking must not be possible for pseudonyms that have not yet been used. It only works if each verifier behaves honestly.

Pooling in AS1: Here, every pseudonym can be pooled because all of them are linkable once the corresponding private user-identifier is known.

Pooling in AS2: Here, only pseudonyms used for AS2.P1b and AS2.P2b are linkable.

6.7 Selected Applications

The following discusses some applications or branches where the schemes presented can be useful to preserve both the interests of user and verifier. The discussions assume that pre- and post-pooling have been implemented.

E-Cash: A typical application for transaction-pseudonyms is e-cash. Each pseudonym is interpreted as a virtual coin (e-coin). The authentication data additionally contains information about the value of a pseudonym (the e-coin) and the user has to pay for each authentication data according to its value. Spending an e-coin implies running the authentication process over the corresponding pseudonym. Since post-pooling is implemented, it can be verified if a user spent an e-coin more than once. If this is the case, his anonymity can be revoked if P2 has been used. Furthermore, if necessary, all his other e-coins can be identified and invalidated. Notice that this is optional. Linking could be carried out if special conditions have been met, such as "the user behaved dishonestly for the third time" or "the user spent some money twice".

Multi-Identity System: The proposed system can also be used to establish role-pseudonyms. For instance, a user could use one pseudonym for auctions, one pseudonym for shopping, one pseudonym for reading the newspaper, etc. In case of illegal actions, the pseudonyms of a suspect could be identified by the revocation centers.

Secure Auctions: The participants can remain anonymous until one wins the auction. In this case, the winner may want to reveal his real identity. If he refuses to pay, the revocation centers can disclose his identity. If specified, other auction chairs could be informed, if the same bidder participates in one of their sessions.

Patent Search: The proposed scheme can be used for research activities in patent databases. Thus, a business rival would be unable to link queries and hence is not able to associate them to a common identifier. Hereby, completely unlinkable pseudonyms are recommended because there is no necessity for optional revocation.

Medical Records in Centralized Databases: If every entry for a medical record is created using a new pseudonym, the privacy of a patient can be completely preserved.

Authenticity of Casino-Chips: Assume that a contactless smart device is embedded in every physical casino-chip. For instance, when a player places a chip in a roulette session, it authenticates itself to the gambling-table immediately. This makes using forged chips detectable. Moreover, the chip can be made traceable or not.

Traceability of Gamblers: Assume that every gambler is provided with a Personal Digital Assistant that is used for online gambling in a casino. When gambling (e.g. roulette), a person authenticates himself using the proposed protocols. If he loses a game he has to pay, otherwise his identity will be revealed for this particular game. Additionally, all the games which he participated in can be linked to him if required. The advantage is that the player is untraceable as long as he behaves honestly.

Tax Fraud and Money Laundering: Suppose one spends money via a pseudonym to conceal the spender's identity. Tax fraud resulting from such a behavior can be prevented by linking the pseudonyms and revoking anonymity.

Service Fraud: A user can register for mobile services. If he does not want to pay for the service, it can be provided for a while as a trial. Later on, the person may attempt to register for the same service again under another pseudonym. Linking the pseudonyms can allow a service vendor to confirm his suspicion.

Insurance Fraud: A user can submit an invoice to multiple insurances for reimbursement under multiple pseudonyms. Linking can help to discover such a fraud.

6.8 Concluding Remarks

This chapter presented two authentication schemes. Both schemes provide protocols to protect the privacy of the authenticator (anonymity and unlinkability) while preserving the interests of the verifier (optional revocation). For practical use, the implementation of AS2 is recommended as it is an advancement of AS1, i.e. is more efficient regarding computational time and space. Thus, it can be even run on smartcards.

An open problem is to design a scheme where the user obtains a root-pseudonym from which he can derive a set of system-wide unique pseudonyms which provably provides optional revocation mechanisms. The advantage of such a scheme would be that only one pseudonym needs to be stored instead of a set of authentication data. Schemes such as [CL01] realize similar ideas but currently do not take the requirement of uniqueness into account. Another open problem is to share the pseudonym certification authority among a set of players, such that a tamper-resistant device is no longer necessary.

AS1 has been implemented in Java by Gerd Oberlechner. A description of the corresponding prototype can be found in [Obe06]. It would be interesting to extend this implementation in order to realize AS2 for some of the smartcard-based applications.

CONCLUSION AND FUTURE WORK

This work dealt with the problem of locally generating system-wide unique numbers which provide a high degree of randomness whilst preserving the privacy of the generating party. Besides giving efficient constructions, it introduced the concept of fusion which can be used to speed up discrete-logarithm-based cryptosystems in particular situations where exponents are longer than necessary from security point of view. A further topic in this work was the generation of shares of a system-wide unique (virtual) secret. The challenge was to never reconstruct the secret during its generation and use. In this context the concept of fusion once again brought improvements. It allowed for constructing efficient protocols in terms of communication complexity and it enabled clean security proofs. Finally, two authentication schemes were presented which have been designed to preserve the interests of both the authenticator and the verifier. This means that the privacy of the authenticator is preserved as long as he behaves honestly. In case of emergency or if the authenticator behaves dishonestly, a revocation center may be able to revoke privacy. The second of the two authentication schemes is efficient enough to be implemented and run on smartcards.

The following discusses several open tasks which would enable a lot of future work.

7.1 Statistical Tests

Some of the practical collision-free number generators given in Section 3.5 have been analyzed with the help of the NIST statistical test suite [NIS01b]. It must be point out that only brief tests were performed since a more detailed analysis would have gone beyond the scope of this work. An important next step would be to perform a detailed analysis of the given and future variants of collision-free number generators.

7.2 Implementation

Some parts of this work have been implemented as prototypes. Thus, a lot of tasks remain open for future work. Some of them are given in the following.

Collision-Free Number Generators: A simple prototype has been implemented for JavaCards [SS07]. This prototype needs to be improved so that it provides a large number of flexible collision-free number generators. Moreover, it would be viable to include an implementation in the Java-Libraries used on ordinary computers.

The Concept of Fusion: Since computations in $\mathbb{G}_p$ and $\mathbb{F}_p$ are expressed through computations in $\mathbb{G}_q$ and $\mathbb{Z}_q$, it is evident to build, for instance, a Java-library.

Collision-Free Distributed Key Generation: A prototype implementation for basic multi-party protocols and discrete-logarithm-based distributed key generation exists [Obe06]. Furthermore, a communication platform has been developed which players can use to send or broadcast messages. It remains to implement the protocols described in Chapter 5 on top of these building blocks.

Unlinkable Anonymous Authentication: AS1 has already been implemented [Obe06]. This implementation could be adapted to realize AS2. Furthermore, the second scheme can be implemented on JavaCards since it requires very little resources.

7.3 Open Problems

This work leaves several problems open to future activities, given in the following.

Collision-Free Number Generators: So far, privacy and randomness have been *argued* to be fulfilled. It remains to *prove* privacy-protection with respect to an unsolved problem. Moreover, one could develop a collision-free number generator whose output is provably polynomially indistinguishable from a true random bit-sequence.

The Concept of Fusion: Chapter 4 introduced the concept of fusion and provided basic problem definitions, security reductions and relations. It is left open for future research to define further problems, reductions and relations with respect to fusion. An important problem that needs to be solved is to find out if the Decision Diffie-Hellman Problem (DDP) and the Fusion Decision Diffie-Hellman Problem (FDDP) are computationally equivalent. This open problem yields an interesting conjecture: if the computationally equivalence between the DDP and FDDP cannot be shown, then the FDDP seems to be a stronger problem than the DDP. Thus, if the DDP is efficiently solved directly (i.e. without solving the Discrete Logarithm Problem or the Diffie-Hellman Problem), then related cryptosystems like ElGamal or Cramer-Shoup will become vulnerable. The latter, for instance, would not be secure against chosen ciphertext attacks anymore. However, if our conjecture remains unrefuted, then such cryptosystems will still remain secure within the fusion-setting. Another interesting topic could be to analyze the security of the RSA cryptosystem in the fusion-setting.

Collision-Free Distributed Key Generation: The protocols given in Chapter 5 enable distributed generation of unique secret ordinary exponents and fusion-exponents. Unfortunately, the protocols for ordinary exponents are inefficient and only provide a bad probability distribution. An open problem is to give a protocol by which a set of players can jointly generate shares of a secret that lies in $\mathbb{Z}_q$, is system-wide unique and provides sufficient randomness to keep computing discrete logarithms hard. Another open problem is to design a protocol by which a system-wide unique RSA modulus can be established without reconstruction of the associated secret parameters.

Unlinkable Anonymous Authentication: Some of the protocols given in Chapter 6 are very efficient, but still for every run of an authentication protocol the user has to use a new pseudonym issued by a trusted party. This has the disadvantage that a lot of storage is necessary. A solution to this problem could be the design of a collision-free pseudonym generator with respect to the design principles of collision-free number generation, but whose output is *provably* revocable by a revocation center. The challenge is that only one initial communication with a trusted party should be used to establish a root-pseudonym.

7.4 Further Applications

In this work several applications have been given for each technique developed. It is left open for several research communities (not necessarily security or cryptography communities) to find several further applications and adapt the presented techniques.

Collision-Free Number Generators: The main-goal was to design collision-free number generators for general purpose. Basically, the ideas have come from the security field, but could be applied to several other areas where the unique identification of entities is important. Hence, they could be also used to achieve the safety of a system.

The Concept of Fusion: So far, it has been shown that the concept of fusion *may* increase the speed of the computations and is useful for collision-free distributed key generation. Since fusion has particular properties, there may be several applications which can take advantage of them for totally different purposes.

Symbols and Notations

The protocols and figures are self-explanatory and require no extra descriptions. Symbols and notations are generally defined whenever they are first used. In any case, the following provides a list of commonly used notations in no particular order.

l_x	bit-length of an integer x
$l_{p_0'}$	lower security bound for a prime modulus p' to make the discrete logarithm problem hard
l_{q_0}	lower security bound for a prime order q to make the discrete logarithm problem hard
$\{0,1\}^*$	set of all words over the alphabet $\{0,1\}$ of arbitrary length
$\{0,1\}^{l_x}$	set of all words over the alphabet $\{0,1\}$ of length l_x
$\{X_i\}_{i \in M}$	set of all X_i where $i \in M$
$\{X_i\}_{i=1}^n$	set of all X_i where $1 \le i \le n$
$\mathcal{H}, \mathcal{G}$	cryptographic hash functions (or random oracles)
$\mathcal{O}_P$	oracle that solves the problem P in poly-time
$\mathcal{P}$	set of n players
$\mathcal{Q}$	set of honest players, i.e. $\mathcal{Q} \subseteq \mathcal{P}$ and $|\mathcal{Q}| > t$
$\mathcal{R}$	set of revocation centers
$\mathcal{T}$	index set $\{0, \ldots, t\}$ of a polynomial's coefficients
$I(\mathcal{P})$	index set of the set $\mathcal{P}$
$\mathbb{G}_p$	group of prime order p (or direct product $\mathbb{G}_q \times \mathbb{G}_q$)
$\mathbb{G}_q$	group of prime order q
$E(\mathbb{Z}_p)$	elliptic curve group where p is prime
$\mathbb{F}_p$	field of order $p = q^2$
$R_{\hat{n}}$	commutative unitary ring of order $\hat{n} = \tilde{n}^2$
$\mathbb{G}_{\hat{n}}$	direct product $QR_n \times QR_n$
QR_q	set of quadratic residues modulo q
QNR_q	set of quadratic non-residues modulo q
QR_n	set of quadratic residues modulo n
ζ	ordinary exponentiation
$(a,b)^{(c,d)}, \xi$	fusion-exponentiation

$fdlog_{\mathbf{g}}(\mathbf{y})$	unique fusion discrete logarithm of $\mathbf{y}$ to the base $\mathbf{g}$
$dlog_g(y)$	unique discrete logarithm of y to the base g
$ecdlog_P(H)$	unique elliptic curve discrete logarithm of H to the base P
$\approx$	approximately
$\equiv$	congruent
$\in_R$	chosen at random
$\|\|$	concatenation
$[a_{i1}\|\|a_{i2}]_{i \in B}$	concatenation of all blocks of the form $a_{i1}\|\|a_{i2}$ where $i \in B$
$\lceil x \rceil$	ceiling of a real number x
$\lfloor x \rfloor$	floor of a real number x
$\overset{?}{=}$	equality check or verification
$ord(a)$	order of the element a
$\tau(a)$	bit-permutation or bit-permuted expansion over one block a
$\pi(a,b)$	bit-permutation or bit-permuted expansion over a and b
φ	Euler's totient function
gcd	greatest common divisor
lcm	least common multiplier
A $\leq_p$ B	poly-reduction from problem A to problem B
A $\equiv_p$ B	abbreviation of A $\leq_p$ B $\wedge$ B $\leq_p$ A
S	signature generation function
V	verification of non-interactive Σ-proof or signature
E_A	asymmetric encryption function
E_S	symmetric encryption function
E	encryption function
f	uniqueness randomization or deterministic one-way function
g	privacy protection function
ASi	authentication scheme $i \in \{1,2\}$
NG	number generator (random, quasi-random or pseudo-random)
Pj	protocol $j \in \{1, 1a, 1b, 2, 2a, 2b, 3\}$
Pre	pre-processor
UG	uniqueness generator
SM	scalar multiplication
EXP	exponentiation
Pr	probability
RC_i	i-th revocation center
Δ	time-interval within which uniqueness is guaranteed
η	pseudonym
Γ	system parameters

Λ	batch of authentication data
λ_i	i-th Lagrange coefficient
ρ	padding up to a bit-length or vector of linking information
ρ'	vector of encrypted linking information
σ	signature
$[s]^q$	polynomial (t-degree) share modulo q
$\langle s \rangle^q$	polynomial (t-degree) fusion-share in $CE(\mathbb{Z}_q)$
ID	public identifier
id	secret identifier
c	counter or challenge
τ	transcript of a non-interactive Σ-proof
P	generator point for $\mathbb{G}_q < E(\mathbb{Z}_p)$
g, g_1, g_2, γ	generators
h	ElGamal public key in $\mathbb{G}_q < \mathbb{Z}_p^*$
H	ElGamal public key in $\mathbb{G}_q < E(\mathbb{Z}_p)$
n	modulus which is the product of two safe-primes p and q
$\widetilde{n}$	hidden order which is a Blum integer
$\widehat{n}$	order in the fusion-setting which is equal to $\widetilde{n}^2$
p', q'	prime numbers (sometimes $p' \equiv q' \equiv 3 \pmod 4$)
p	prime number or equal to q^2
q	prime number
t	number of tolerated adversaries (degree of sharing-polynomial)
pk	public key
sk	signature generation key
vk	verification key
ek	encryption key
dk	decryption key

ACRONYMS

Appendix B

AES	Advanced Encryption Standard
BB	Bulletin Board
CFNG	Collision-Free Number Generator
CT	Ciphertext-Stealing
DDP	Decision Diffie-Hellman Problem
DES	Data Encryption Standard
DHP	Diffie-Hellman Problem
Div-DHP	Divisible Diffie-Hellman Problem
DLP	Discrete Logarithm Problem
FDDP	Full-Fusion Decision Diffie-Hellman Problem
FDHP	Full-Fusion Diffie-Hellman Problem
FDLP	Full-Fusion Discrete Logarithm Problem
HDDP	Half-Fusion Decision Diffie-Hellman Problem
HDHP	Half-Fusion Diffie-Hellman Problem
HDLP	Half-Fusion Discrete Logarithm Problem
IDEA	Ideal Data Encryption Algorithm
IDLP	Interval Discrete Logarithm Problem
MAC	Media Access Control
OAEP	Optimal Asymmetric Encryption Padding
PCA	Pseudonym Certification Authority
PRNG	Pseudo-Random Number Generator
QRNG	Quasi-Random Number Generator
RFC	Request For Comments
Ri	Requirement i
RNG	Random Number Generator
RSA	Rivest Shamir Adleman
TRD	Tamper-Resistant Device
UI	Unique Identifier
UUID	Universally Unique Identifier

REFERENCES

[AABN02] M. Abdalla, J. H. An, M. Bellare, and C. Namprempre. From Identification to Signatures via the Fiat-Shamir Transform: Minimizing Assumptions for Security and Forward-Security. In L. R. Knudsen, editor,
Advances in Cryptology – EUROCRYPT'02, volume 2332 of *Lecture Notes
in Computer Science*, pages 418–433. Springer, 2002.

[AB83] C. A. Asmuth and J. Bloom. A Modular Approach to Key Safeguarding.
In *IEEE Transactions on Information Theory*, volume 29, pages 208–210.
IEEE Press, 1983.

[ACJT00] G. Ateniese, J. Camenisch, M. Joye, and G. Tsudik. A Practical and Provably Secure Coalition-Resistant Group Signature Scheme. In M. Bellare,
editor, *Advances in Cryptology – CRYPTO'00*, volume 1880 of *Lecture
Notes in Computer Science*, pages 255–270. Springer, 2000.

[ACS02] J. Algesheimer, J. Camenisch, and V. Shoup. Efficient Computation
Modulo a Shared Secret with Application to the Generation of Shared
Safe-Prime Products. In M. Yung, editor, *Advances in Cryptology –
CRYPTO'02*, volume 2442 of *Lecture Notes in Computer Science*, pages
417–432. Springer, 2002.

[ADN06] J. F. Almansa, I. Damgård, and J. B. Nielsen. Simplified Threshold RSA
with Adaptive and Proactive Security. In S. Vaudenay, editor, *Advances
in Cryptology – EUROCRYPT'06*, volume 4004 of *Lecture Notes in Computer Science*, pages 593–611. Springer, 2006.

[ANS86] ANSI INCITS 4-1986 (R2002): Information Systems – Coded Character
Sets – 7-Bit American National Standard Code for Information Inter-
change (7-Bit ASCII), 1986.

[BBR99] E. Biham, D. Boneh, and O. Reingold. Breaking Generalized Diffie-
Hellman Modulo a Composite is no Easier than Factoring. *Information
Processing Letters*, 70(2):83–87, 1999.

[BCC88]   G. Brassard, D. Chaum, and C. Crépeau. Minimum Disclosure Proofs of Knowledge. *J. of Computer and System Sciences*, 37(2):156–189, 1988.

[BDS98]   M. Burmester, Y. Desmedt, and J. Seberry. Equitable Key Escrow with Limited Time Span (or, How to Enforce Time Expiration Cryptographically). In K. Ohta and D. Pei, editors, *Advances in Cryptology – ASIACRYPT'98*, volume 1514 of *Lecture Notes in Computer Science*, pages 380–391. Springer, 1998.

[BDZ03]   F. Bao, R. H. Deng, and H. Zhu. Variations of Diffie-Hellman Problem. In S. Qing, D. Gollmann, and J. Zhou, editors, *Proceedings of the 5th International Conference on Information and Communications Security – ICICS'03*, volume 2836 of *Lecture Notes in Computer Science*, pages 301–312. Springer, 2003.

[Bea91]   D. Beaver. Secure Multiparty Protocols and Zero-Knowledge Proof Systems Tolerating a Faulty Minority. *J. of Cryptology*, 4(2):75–122, 1991.

[BF01]   D. Boneh and M. Franklin. Efficient Generation of Shared RSA Keys. *Journal of the ACM*, 48(4):702–722, 2001.

[BFM88]   M. Blum, P. Feldman, and S. Micali. Non-Interactive Zero-Knowledge and Its Applications. In *Proceedings of the 20th Annual ACM Symposium on the Theory of Computing – STOC'88*, pages 103–112. ACM Press, 1988.

[BG93]   M. Bellare and O. Goldreich. On Defining Proofs of Knowledge. In E. F. Brickell, editor, *Advances in Cryptology – CRYPTO'92*, volume 740 of *Lecture Notes in Computer Science*, pages 390–420. Springer, 1993.

[Bla79]   G. R. Blakely. Safeguarding Cryptographic Keys. In *Proceedings of AFIPS 1979*, volume 48, pages 313–317, 1979.

[Ble98]   D. Bleichenbacher. Chosen Ciphertext Attacks Against Protocols Based on the RSA Encryption Standard PKCS #1. In H. Krawczyk, editor, *Advances in Cryptology – CRYPTO'98*, volume 1462 of *Lecture Notes in Computer Science*, pages 1–12. Springer, 1998.

[BOGW88] M. Ben-Or, S. Goldwasser, and A. Wigderson. Completeness Theorems for Non-Cryptographic Fault-Tolerant Distributed Computation. In *Proceedings of the 20th Annual ACM Symposium on the Theory of Computing – STOC'88*, pages 1–10. ACM Press, 1988.

[Bon98]   D. Boneh. The Decision Diffie-Hellman Problem. In J. Buhler, editor, *Proceedings of the Third International Symposium on Algorithmic Number Theory – ANTS-III*, volume 1423 of *Lecture Notes in Computer Science*, pages 48–63. Springer, 1998.

[Bos05]   S. Bosch. *Algebra*. Springer, 2005.

[Bou00]   F. Boudot. Efficient Proofs that a Committed Number Lies in an Interval. In B. Preneel, editor, *Advances in Cryptology – EUROCRYPT'00*, volume 1807 of *Lecture Notes in Computer Science*, pages 431–444. Springer, 2000.

[BPVY00]   E. F. Brickell, D. Pointcheval, S. Vaudenay, and M. Yung. Design Validations for Discrete Logarithm Based Signature Schemes. In H. Imai and Y. Zheng, editors, *Proc. of Public Key Cryptography – PKC'00*, vol. 1751 of *Lecture Notes in Computer Science*, pages 276–292. Springer, 2000.

[BR93]   M. Bellare and P. Rogaway. Random Oracles are Practical: A Paradigm for Designing Efficient Protocols. In *Proceedings of the 1st ACM Conference on Computer and Communications Security – CCS'93*, pages 62–73. ACM Press, 1993.

[BR95]   M. Bellare and P. Rogaway. Optimal Asymmetric Encryption. In A. De Santis, editor, *Advances in Cryptology – EUROCRYPT'94*, volume 950 of *Lecture Notes in Computer Science*, pages 92–111. Springer, 1995.

[BSZ04]   M. Bellare, H. Shi, and C. Zhang. Foundations of Group Signatures: The Case of Dynamic Groups. Crypt. ePrint Archive, Report 2004/077, 2004.

[BT00]   S. R. Blackburn and E. Teske. Baby-Step Giant-Step Algorithms for Non-Uniform Distributions. In W. Bosma, editor, *Proc. of the Fourth International Symposium on Algorithmic Number Theory – ANTS-IV*, volume 1838 of *Lecture Notes in Computer Science*, pages 153–168. Springer, 2000.

[Bun02]   P. Bundschuh. *Einführung in die Zahlentheorie*. Springer, 5th ed., 2002.

[Cac95]   C. Cachin. On-Line Secret Sharing. In C. Boyd, editor, *Proceedings of the 5th IMA Conference on Cryptography and Coding*, volume 1025 of *Lecture Notes in Computer Science*, pages 190–198. Springer, 1995.

[CCD88]   D. Chaum, C. Crépeau, and I. Damgård. Multiparty Unconditionally Secure Protocols. In *Proceedings of the 20th Annual ACM Symposium on the Theory of Computing – STOC'88*, pages 11–19. ACM Press, 1988.

[CD97]   R. Cramer and I. Damgård. Linear Zero-Knowledge – A Note on Efficient
          Zero-Knowledge Proofs and Arguments. In *Proceedings of the 29th Annual
          ACM Symposium on the Theory of Computing – STOC'97*, pages 436–445.
          ACM Press, 1997.

[CD98]   R. Cramer and I. Damgård. Zero-Knowledge Proofs for Finite Field Arith-
          metic; or: Can Zero-Knowledge be for Free? In H. Krawczyk, editor,
          Advances in Cryptology – CRYPTO'98, volume 1462 of *Lecture Notes in
          Computer Science*, pages 424–441. Springer, 1998.

[CDM00]   R. Cramer, I. Damgård, and U. M. Maurer. General Secure Multi-Party
          Computation from any Linear Secret-Sharing Scheme. In B. Preneel, ed-
          itor, *Advances in Cryptology – EUROCRYPT'00*, volume 1807 of *Lecture
          Notes in Computer Science*, pages 316–334. Springer, 2000.

[CE87]   D. Chaum and J.-H. Evertse. A Secure and Privacy-Protecting Protocol
          for Transmitting Personal Information Between Organizations. In A. M.
          Odlyzko, editor, *Advances in Cryptology – CRYPTO'86*, volume 263 of
          Lecture Notes in Computer Science, pages 118–167. Springer, 1987.

[CEG87]   D. Chaum, J.-H. Evertse, and J. van de Graaf. An Improved Protocol
          for Demonstrating Possession of Discrete Logarithms and Some General-
          izations. In D. Chaum and W. L. Price, editors, *Advances in Cryptology
          – EUROCRYPT'87*, volume 304 of *Lecture Notes in Computer Science*,
          pages 127–141. Springer, 1987.

[CG98]   D. Catalano and R. Gennaro. New Efficient and Secure Protocols for
          Verifiable Signature Sharing and Other Applications. In H. Krawczyk,
          editor, *Advances in Cryptology – CRYPTO'98*, volume 1462 of *Lecture
          Notes in Computer Science*, pages 105–120. Springer, 1998.

[CG99]   R. Canetti and S. Goldwasser. An Efficient Threshold Public Key Cryp-
          tosystem Secure Against Adaptive Chosen Ciphertext Attack. In J. Stern,
          editor, *Advances in Cryptology – EUROCRYPT'99*, volume 1592 of *Lec-
          ture Notes in Computer Science*, pages 90–106. Springer, 1999.

[CGJ+99] R. Canetti, R. Gennaro, S. Jarecki, H. Krawczyk, and T. Rabin. Adaptive
          Security for Threshold Cryptosystems. In M. J. Wiener, editor, *Advances
          in Cryptology – CRYPTO'99*, volume 1666 of *Lecture Notes in Computer
          Science*, pages 98–115. Springer, 1999.

[CGMA85] B. Chor, S. Goldwasser, S. Micali, and B. Awerbuch. Verifiable Secret Sharing and Achieving Simultaneity in the Presence of Faults. In *Proceedings of the 26th Annual IEEE Symposium on Foundations of Computer Science – FOCS'85*, pages 383–395. IEEE Press, 1985.

[CGS97]   R. Cramer, R. Gennaro, and B. Schoenmakers. A Secure and Optimally Efficient Multi-Authority Election Scheme. In W. Fumy, editor, *Advances in Cryptology – EUROCRYPT'97*, volume 1233 of *Lecture Notes in Computer Science*, pages 103–118. Springer, 1997.

[Cha81]   D. L. Chaum. Untraceable Electronic Mail, Return Addresses, and Digital Pseudonyms. *Communications of the ACM*, 24(2):84–90, 1981.

[CHLS97] T. Collins, D. Hopkins, S. Langford, and M. Sabin. Public Key Cryptographic Apparatus and Method. US Patent #5,848,159, 1997.

[CL01]   J. Camenisch and A. Lysyanskaya. An Efficient System for Nontransferable Anonymous Credentials with Optional Anonymity Revocation. In B. Pfitzmann, editor, *Advances in Cryptology – EUROCRYPT'01*, volume 2045 of *Lecture Notes in Computer Science*, pages 93–118. Springer, 2001.

[Coc97]   C. Cocks. Split Knowledge Generation of RSA Parameters. In M. Darnell, editor, *Proceedings of the 6th IMA International Conference on Cryptography and Coding*, volume 1355 of *Lecture Notes in Computer Science*, pages 89–95. Springer, 1997.

[Col]   ColdFusion. http://kb.adobe.com/selfservice/viewContent.do? externalId=tn_18133&sliceId=2.

[CP93]   D. Chaum and T. Pedersen. Wallet Databases with Observers. In E. F. Brickell, editor, *Advances in Cryptology – CRYPTO'92*, volume 740 of *Lecture Notes in Computer Science*, pages 89–105. Springer, 1993.

[CPS95]   J. Camenisch, J.-M. Piveteau, and M. Stadler. Blind Signatures Based on the Discrete Logarithm Problem. In A. De Santis, editor, *Advances in Cryptology – EUROCRYPT'94*, volume 950 of *Lecture Notes in Computer Science*, pages 428–432. Springer, 1995.

[Cra96]   R. Cramer. *Modular Design of Secure yet Practical Cryptographic Protocols*. PhD thesis, University of Amsterdam, 1996.

[CS97]   J. Camenisch and M. Stadler. Efficient Group Signature Schemes for
           Large Groups (Extended Abstract). In B. S. Kaliski Jr., editor, *Advances
           in Cryptology – CRYPTO'97*, volume 1294 of *Lecture Notes in Computer
           Science*, pages 410–424. Springer, 1997.

[CS98]   R. Cramer and V. Shoup. A Practical Public Key Cryptosystem Provably
           Secure Against Adaptive Chosen Ciphertext Attack. In H. Krawczyk,
           editor, *Advances in Cryptology – CRYPTO'98*, volume 1462 of *Lecture
           Notes in Computer Science*, pages 13–25. Springer, 1998.

[CS99]   R. Cramer and V. Shoup. Signature Schemes Based on the Strong RSA
           Assumption. In *Proceedings of the 6th ACM Conference on Computer and
           Communications Security – CCS'99*, pages 46–51. ACM Press, 1999.

[CS04]   R. Cramer and V. Shoup. Design and Analysis of Practical Public-Key
           Encryption Schemes Secure against Adaptive Chosen Ciphertext Attack.
           SIAM Journal on Computing, 33(1):167–226, 2004.

[CvH91]   D. Chaum and E. van Heyst. Group Signatures. In D. W. Davies, editor,
           Advances in Cryptology – EUROCRYPT'91, volume 547 of *Lecture Notes
           in Computer Science*, pages 257–265. Springer, 1991.

[Dam99]   I. Damgård. Commitment Schemes and Zero-Knowledge Protocols. In
           *Lectures on Data Security, Modern Cryptology in Theory and Practice,
           Summer School, Aarhus, Denmark, July 1998*, volume 1561 of *Lecture
           Notes in Computer Science*, pages 63–86. Springer, 1999.

[Dam00]   I. Damgård. Efficient Concurrent Zero-Knowledge in the Auxiliary String
           Model. In B. Preneel, editor, *Advances in Cryptology – EUROCRYPT'00*,
           volume 1807 of *Lecture Notes in Computer Science*, pages 418–430.
           Springer, 2000.

[DAN02]   P. Drayton, B. Albahari, et al. *C# in a Nutshell*. O'Reilly, 2002.

[DD05]   I. Damgård and K. Dupont. Efficient Threshold RSA Signatures with
           General Moduli and No Extra Assumptions. In S. Vaudenay, editor, *Pro-
           ceedings of Public Key Cryptography – PKC'05*, volume 3386 of *Lecture
           Notes in Computer Science*, pages 346–361. Springer, 2005.

[DDN91]   D. Dolev, C. Dwork, and M. Naor. Non-Malleable Cryptography. In *Pro-
           ceedings of the 23rd Annual ACM Symposium on the Theory of Computing
           – STOC'91*, pages 542–552. ACM Press, 1991.

[DeL84]   J. M. DeLaurentis. A Further Weakness in the Common Modulus Protocol in the RSA Cryptoalgorithm. *Cryptologia*, 8(3):253–259, 1984.

[Des88]   Y. Desmedt. Society and Group Oriented Cryptography: A New Concept. In C. Pomerance, editor, *Advances in Cryptology – CRYPTO'87*, volume 293 of *Lecture Notes in Computer Science*, pages 120–127. Springer, 1988.

[Des98]   Y. Desmedt. Some Recent Research Aspects of Threshold Cryptography. In E. Okamoto, G. I. Davida, and M. Mambo, editors, *Proceedings of the First International Workshop on Information Security – ISW'97*, volume 1396 of *Lecture Notes in Computer Science*, pages 158–173. Springer, 1998.

[DF90]   Y. Desmedt and Y. Frankel. Threshold Cryptosystems. In G. Brassard, editor, *Advances in Cryptology – CRYPTO'89*, volume 435 of *Lecture Notes in Computer Science*, pages 307–315. Springer, 1990.

[DF92]   Y. Desmedt and Y. Frankel. Shared Generation of Authenticators and Signatures (Extended Abstract). In J. Feigenbaum, editor, *Advances in Cryptology – CRYPTO'91*, volume 576 of *Lecture Notes in Computer Science*, pages 457–469. Springer, 1992.

[DH76]   W. Diffie and M. E. Hellman. New Directions in Cryptography. *IEEE Transactions on Information Theory*, IT-22(6):644–654, 1976.

[DJ97]   Y. Desmedt and S. Jajodia. Redistributing Secret Shares to New Access Structures and Its Applications. Technical Report ISSE TR-97-01, George Mason University, 1997.

[DM84]   D. Dorninger and W. Müller. *Allgemeine Algebra und Anwendungen*. Teubner, 1984.

[DNS98]   C. Dwork, M. Naor, and A. Sahai. Concurrent Zero-Knowledge. In *Proceedings of the 30th Annual ACM Symposium on the Theory of Computing – STOC'98*, pages 409–418. ACM Press, 1998.

[ElG85a]   T. ElGamal. A Public Key Cryptosystem and a Signature Scheme Based on Discrete Logarithms. In G. R. Blakley and D. Chaum, editors, *Advances in Cryptology – CRYPTO'84*, volume 196 of *Lecture Notes in Computer Science*, pages 10–18. Springer, 1985.

[ElG85b]   T. ElGamal. A Public Key Cryptosystem and a Signature Scheme Based on Discrete Logarithms. *IEEE Transactions on Information Theory*, 31(4):469–472, 1985.

[Fel87]   P. Feldman. A Practical Scheme for Non-Interactive Verifiable Secret Sharing. In *IEEE Symposium on Foundations of Computer Science*, pages 427–437. IEEE Press, 1987.

[FGMY97] Y. Frankel, P. Gemmell, P. D. MacKenzie, and M. Yung. Proactive RSA. In B. S. Kaliski Jr., editor, *Advances in Cryptology – CRYPTO'97*, volume 1294 of *Lecture Notes in Computer Science*, pages 440–454. Springer, 1997.

[Fin99]   K. Finkenzeller. *RFID Handbook*. John Wiley & Sons, 1999.

[FMY98]   Y. Frankel, P. D. MacKenzie, and M. Yung. Robust Efficient Distributed RSA-Key Generation. In *Proceedings of the 17th Annual ACM Symposium on Principles of Distributed Computing – PODC'98*, pages 663–672. ACM Press, 1998.

[FO97]   E. Fujisaki and T. Okamoto. Statistical Zero Knowledge Protocols to Prove Modular Polynomial Relations. In B. S. Kaliski Jr., editor, *Advances in Cryptology – CRYPTO'97*, volume 1294 of *Lecture Notes in Computer Science*, pages 16–30. Springer, 1997.

[FPS01]   P.-A. Fouque, G. Poupard, and J. Stern. Sharing Decryption in the Context of Voting or Lotteries. In Y. Frankel, editor, *Proceedings of Financial Cryptography – FC'00*, volume 1962 of *Lecture Notes in Computer Science*, pages 90–104. Springer, 2001.

[FS87]   A. Fiat and A. Shamir. How to Prove Yourself: Practical Solutions to Identification and Signature Problems. In A. M. Odlyzko, editor, *Advances in Cryptology – CRYPTO'86*, volume 263 of *Lecture Notes in Computer Science*, pages 186–194. Springer, 1987.

[FTY96]   Y. Frankel, Y. Tsiounis, and M. Yung. "Indirect Discourse Proof": Achieving Efficient Fair Off-Line E-Cash. In K. Kim and T. Matsumoto, editors, *Advances in Cryptology – ASIACRYPT'96*, volume 1163 of *Lecture Notes in Computer Science*, pages 286–300. Springer, 1996.

[GGM86]   O. Goldreich, S. Goldwasser, and S. Micali. How to Construct Random Functions. *Journal of the ACM*, 33(4):792–807, 1986.

[Gir91]   M. Girault. An Identity-Based Identification Scheme Based on Discrete Logarithms Modulo a Composite Number. In I. Damgård, editor, *Advances in Cryptology – EUROCRYPT'90*, volume 473 of *Lecture Notes in Computer Science*, pages 481–486. Springer, 1991.

[GJJS04] P. Golle, M. Jakobsson, A. Juels, and P. F. Syverson. Universal Re-Encryption for Mixnets. In T. Okamoto, editor, *The Cryptographers' Track at the RSA Conference 2004 – CT-RSA'04*, volume 2964 of *Lecture Notes in Computer Science*, pages 163–178. Springer, 2004.

[GJKR96] R. Gennaro, S. Jarecki, H. Krawczyk, and T. Rabin. Robust Threshold DSS Signatures. In U. M. Maurer, editor, *Advances in Cryptology – EUROCRYPT'96*, volume 1070 of *Lecture Notes in Computer Science*, pages 354–371. Springer, 1996.

[GJKR99] R. Gennaro, S. Jarecki, H. Krawczyk, and T. Rabin. Secure Distributed Key Generation for Discrete-Log Based Cryptosystems. In J. Stern, editor, *Advances in Cryptology – EUROCRYPT'99*, volume 1592 of *Lecture Notes in Computer Science*, pages 295–310. Springer, 1999.

[GJKR03] R. Gennaro, S. Jarecki, H. Krawczyk, and T. Rabin. Secure Applications of Pedersen's Distributed Key Generation Protocol. In M. Joye, editor, *The Cryptographers' Track of the RSA Conference 2003 – CT-RSA'03*, volume 2612 of *Lecture Notes in Computer Science*, pages 373–390. Springer, 2003.

[GJKR07] R. Gennaro, S. Jarecki, H. Krawczyk, and T. Rabin. Secure Distributed Key Generation for Discrete-Log Based Cryptosystems. *Journal of Cryptology*, 20(1):51–83, 2007.

[GK03] S. Goldwasser and Y. T. Kalai. On the (In)security of the Fiat-Shamir Paradigm. In *Proc. of the 44th Annual IEEE Symposium on Foundations of Computer Science – FOCS'03*, pages 102–113. IEEE Press, 2003.

[GM82] S. Goldwasser and S. Micali. Probabilistic Encryption & How to Play Mental Poker Keeping Secret all Partial Information. In *Proceedings of the 14th Annual ACM Symposium on the Theory of Computing – STOC'82*, pages 365–377. ACM Press, 1982.

[GM84] S. Goldwasser and S. Micali. Probabilistic Encryption. *Journal of Computer and System Sciences*, 28(2):270–299, 1984.

[GMR89] S. Goldwasser, S. Micali, and C. Rackoff. The Knowledge Complexity of Interactive Proof Systems. *SIAM J. on Computing*, 18(1):186–208, 1989.

[GMW87] O. Goldreich, S. Micali, and A. Wigderson. How to Play ANY Mental Game. In *Proceedings of the 19th Annual ACM Symposium on the Theory of Computing – STOC'87*, pages 218–229. ACM Press, 1987.

[Gol02]   O. Goldreich. Zero-Knowledge Twenty Years After Its Invention. Tech-
            nical report, Department of Computer Science and Applied Mathematics,
            Weizmann Institute of Science, Israel, 2002.

[GRR98]   R. Gennaro, M. O. Rabin, and T. Rabin. Simplified VSS and Fast-Track
            Multiparty Computations with Applications to Threshold Cryptography.
            In *Proceedings of the 17th Annual ACM Symposium on Principles of Dis-
            tributed Computing – PODC'98*, pages 101–111. ACM Press, 1998.

[GRS96]   D. M. Goldschlag, M. G. Reed, and P. F. Syverson. Hiding Routing Infor-
            mation. In R. J. Anderson, editor, *Proceedings of the First International
            Workshop on Information Hiding*, volume 1174 of *Lecture Notes in Com-
            puter Science*, pages 137–150. Springer, 1996.

[GSV98]   O. Goldreich, A. Sahai, and S. Vadhan. Honest-Verifier Statistical Zero-
            Knowledge Equals General Statistical Zero-Knowledge. In *Proceedings
            of the 30th Annual ACM Symposium on the Theory of Computing –
            STOC'98*, pages 399–408. ACM Press, 1998.

[GV88]   O. Goldreich and R. Vainish. How to Solve any Protocol Problem – An
            Efficiency Improvement. In C. Pomerance, editor, *Advances in Cryptology
            – CRYPTO'87*, volume 293 of *Lecture Notes in Computer Science*, pages
            73–86. Springer, 1988.

[Har94]   L. Harn. Group-Oriented (t, n) Threshold Digital Signature Scheme and
            Multisiganture. In *IEEE Proceedings – Computers and Digital Techniques*,
            volume 141, pages 307–313. IEEE Press, 1994.

[Hir01]   M. Hirt. *Multi-Party Computation: Efficient Protocols, General Adver-
            saries, and Voting*. PhD thesis, ETH Zurich, 2001. Reprint as vol. 3
            of *ETH Series in Information Security and Cryptography*, ISBN 3-89649-
            747-2, Hartung-Gorre Verlag, Konstanz, 2001.

[HJKY95]   A. Herzberg, S. Jarecki, H. Krawczyk, and M. Yung. Proactive Secret
            Sharing, Or: How to Cope With Perpetual Leakage. In D. Coppersmith,
            editor, *Advances in Cryptology – CRYPTO'95*, volume 963 of *Lecture
            Notes in Computer Science*, pages 339–352. Springer, 1995.

[HM00]   M. Hirt and U. Maurer. Player Simulation and General Adversary Struc-
            tures in Perfect Multiparty Computation. *J. of Crypt.*, 13(1):31–60, 2000.

[HMP94]   P. Horster, M. Michels, and H. Petersen. Generalized Blind Signature
            Schemes Based on the Discrete Logarithm Problem. Technical Report
            TR-94-7, University of Technology Chemnitz-Zwickau, 1994.

[HMV04]   D. Hankerson, A. J. Menezes, and S. A. Vanstone. *Guide to Elliptic Curve
            Cryptography*. Springer, 2004.

[Hor98]   P. Horster. Dublettenfreie Schlüsselgenerierung durch isolierte Instanzen.
            In *Chipkarten*, pages 104–119. Vieweg Verlag, 1998.

[HPM94]   P. Horster, H. Petersen, and M. Michels. Meta-ElGamal Signature
            Schemes. In *Proceedings of the 2nd ACM Conference on Computer and
            Communications Security – CCS'94*, pages 96–107. ACM Press, 1994.

[HS98]   P. Horster and P. Schartner. Bemerkungen zur Erzeugung dublettenfreier
            Primzahlen. In Patrick Horster, editor, *Proceedings of Sicherheitsinfras-
            trukturen*, pages 358–368. Vieweg Verlag, 1998.

[HSW98a]   P. Horster, P. Schartner, and P. Wohlmacher. Key Management. In
            G. Papp and R. Posch, editors, *Proceedings of the IFIP TC11 14th Inter-
            national Information Security Conference – SEC'98*, pages 37–48, 1998.

[HSW98b]   P. Horster, P. Schartner, and P. Wohlmacher. Special Aspects of Key
            Generation. In *Proceedings of Information Technologies in Science, En-
            gineering, Education, Health Service*, pages 345–350. Printed Scientific
            Works, 1998.

[HTY99]   H. Handschuh, Y. Tsiounis, and M. Yung. Decision Oracles are Equiva-
            lent to Matching Oracles. In H. Imai and Y. Zheng, editors, *Proceedings
            of Public Key Cryptography – PKC'99*, volume 1560 of *Lecture Notes in
            Computer Science*, pages 276–289. Springer, 1999.

[IEE00]   IEEE P1363 / D1 (Draft Version 1) Standard Specifications For Public
            Key Cryptography. ANSI/IEEE, 2000.

[IPC02]   IPCOM. Method of Generating Unique Quasi-Random Numbers as a
            Function of Time and Space. PriorArtDB, IPCOM#000007118D, 2002.

[ISO96]   ISO/IEC 11578:1996: Remote Procedure Call (RPC), 1996.

[ISO05]   ISO/IEC 9834-8:2005: Generation and Registration of Universally Unique
            Identifiers (UUIDs) and their use as ASN.1 Obj. Ident. Components, 2005.

[IT04]   ITU-T X.667: Information Technology – Open Systems Interconnection – Procedures for the Operation of OSI Registration Authorities: Generation and Registration of Universally Unique Identifiers (UUIDs) and their Use as ASN.1 Object Identifier Components, 2004.

[Java]   Java. Platform Standard Edition (J2SE) 5.0. http://java.sun.com/ j2se/1.5.0/docs/api/java/util/UUID.html.

[Javb]   Javascript. http://www.af-design.com/services/javascript/uuid/.

[JS05]   S. Jarecki and N. Saxena. Further Simplifications in Proactive RSA Signatures. In J. Kilian, editor, *Proceedings of the 2nd Theory of Cryptography Conference – TCC'05*, volume 3378 of *Lecture Notes in Computer Science*, pages 510–528. Springer, 2005.

[JY96]   M. Jakobsson and M. Yung. Revokable and Versatile Electronic Money. In *Proceedings of the 3rd ACM Conference on Computer and Communications Security – CCS'96*, pages 76–87. ACM Press, 1996.

[KCKB03]   J. Kim, S. Choi, K. Kim, and C. Boyd. Anonymous Authentication Protocol for Dynamic Groups with Power-Limited Devices. In *Proceedings of Symposium on Cryptography and Information Security – SCIS'03*, number 2, pages 405–410, 2003.

[Kee03]   KeeLoq. Code Hopping Encoder. HCS201 Datasheet, April 2003.

[KGH83]   E. Karnin, J. Greene, and M. Hellman. On Secret Sharing Systems. In *IEEE Transactions on Information Theory*, volume 29, pages 35–41. IEEE Press, 1983.

[Kil01]   E. Kiltz. A Tool Box of Cryptographic Functions Related to the Diffie-Hellman Function. In C. P. Rangan and C. Ding, editors, *Progress in Cryptology – INDOCRYPT'01*, volume 2247 of *Lecture Notes in Computer Science*, pages 339–350. Springer, 2001.

[Kle01]   J. Klensin. RFC2821: Simple Mail Tranfer Protocol. http://www.ietf. org/rfc/rfc2821.txt, 2001.

[KRJ98]   D. Kesdogan, P. Reichl, and K. Junghärtchen. Distributed Temporary Pseudonyms: A New Approach for Protecting Location Information in Mobile Communication Networks. In J.-J. Quisquater, Y. Deswarte, C. Meadows, and D. Gollmann, editors, *Proceedings of the 5th European Symposium on Research in Computer Security – ESORICS'98*, volume 1485 of *Lecture Notes in Computer Science*, pages 295–312. Springer, 1998.

[KTY04]   A. Kiayias, Y. Tsiounis, and M. Yung. Traceable Signatures. Cryptology
          ePrint Archive, Report 2004/007, 2004.

[Lin03]   Y. Lindell. *Composition of Secure Multi-Party Protocols: A Comprehen-
          sive Study*, vol. 2815 of *Lecture Notes in Comp. Science*. Springer, 2003.

[LM91]   X. Lai and J. L. Massey. A Proposal for a New Block Encryption Stan-
          dard. In I. Damgård, editor, *Advances in Cryptology – EUROCRYPT'90*,
          vol. 473 of *Lecture Notes in Comp. Science*, pages 389–404. Springer, 1991.

[LNV99]   T. Van Le, K. Q. Nguyen, and V. Varadharajan. How to Prove that a
          Committed Number is Prime. In K.-Y. Lam, E. Okamoto, and C. Xing,
          editors, *Advances in Cryptology – ASIACRYPT'99*, volume 1716 of *Lec-
          ture Notes in Computer Science*, pages 208–218. Springer, 1999.

[LS91]   D. Lapidot and A. Shamir. Publicly Verifiable Non-Interactive Zero-
          Knowledge Proofs. In A. Menezes and S. A. Vanstone, editors, *Advances
          in Cryptology – CRYPTO'90*, volume 537 of *Lecture Notes in Computer
          Science*, pages 353–365. Springer, 1991.

[LSA01]   E. Limpert, W. A. Stahel, and M. Abbt. Log-Normal Distributions Across
          the Sciences: Keys and Clues. *BioScience*, 51(5):341–352, May 2001.

[Mao98]   W. Mao. Guaranteed Correct Sharing of Integer Factorization with Off-
          Line Shareholders. In H. Imai and Y. Zheng, editors, *Proceedings of Public
          Key Cryptography – PKC'98*, volume 1431 of *Lecture Notes in Computer
          Science*, pages 60–71. Springer, 1998.

[Mau94]   U. M. Maurer. Towards the Equivalence of Breaking the Diffie-Hellman
          Protocol and Computing Discrete Logarithms. In Y. Desmedt, editor,
          Advances in Cryptology – CRYPTO'94, volume 839 of *Lecture Notes in
          Computer Science*, pages 271–281. Springer, 1994.

[McC90]   K. S. McCurley. The Discrete Logarithm Problem. In C. Pomerance,
          editor, *Cryptology and Computational Number Theory*, volume 42, pages
          49–74. American Mathematical Society, 1990.

[Mil86]   V. S. Miller. Use of Elliptic Curves in Cryptography. In H. C. Williams, ed-
          itor, *Advances in Cryptology – CRYPTO'85*, volume 218 of *Lecture Notes
          in Computer Science*, pages 417–426. Springer, 1986.

[MM82]   C. G. Meyer and S. M. Matyas. *Cryptography: A New Dimension in
          Computer Data Security*. John Wiley & Sons, 1982.

[MR02]   S. Micali and L. Reyzin. Improving the Exact Security of Digital Signature
           Schemes. *Journal of Cryptology*, 15(1):1–18, 2002.

[MS05]   M. Mealling, R. Salz. RFC4122: A Universally Unique IDentifier (UUID)
           URN Namespace. http://www.ietf.org/rfc/rfc4122.txt, 2005.

[MVO96]   A. J. Menezes, S. A. Vanstone, and P. C. Van Oorschot. *Handbook of
           Applied Cryptography*. CRC Press, 1996.

[MW96]   U. M. Maurer and S. Wolf. Diffie-Hellman Oracles. In N. Koblitz, editor,
           Advances in Cryptology – CRYPTO'96, volume 1109 of *Lecture Notes in
           Computer Science*, pages 268–282. Springer, 1996.

[MW98a]   U. M. Maurer and S. Wolf. Diffie-Hellman, Decision Diffie-Hellman, and
           Discrete Logarithms. In *Proceedings of ISIT'98*, page 327. IEEE Informa-
           tion Theory Society, August 1998.

[MW98b]   U. M. Maurer and S. Wolf. Lower Bounds on Generic Algorithms
           in Groups. In K. Nyberg, editor, *Advances in Cryptology – EURO-
           CRYPT'98*, volume 1403 of *Lecture Notes in Computer Science*, pages
           72–84. Springer, 1998.

[MW99]   U. Maurer and S. Wolf. The Relationship Between Breaking the Diffie-
           Hellman Protocol and Computing Discrete Logarithms. *SIAM Journal on
           Computing*, 28(5):1689–1721, 1999.

[MW00]   U. M. Maurer and S. Wolf. The Diffie-Hellman Protocol. *Designs, Codes,
           and Cryptography*, 19(3):147–171, Jan 2000.

[MyS]   MySQL. http://dev.mysql.com/doc/refman/4.1/en/miscellaneous-
           functions.html.

[Nie92]   H. Niederreiter. *Random Number Generation and Quasi-Monte Carlo
           Methods*. Society for Industrial and Applied Mathematics, 1992.

[NIS80]   NIST FIPS Publication 81: DES Modes of Operation, 1980.

[NIS94]   NIST FIPS Publication 186: Digital Signature Standard, 1994.

[NIS98]   NIST SKIPJACK and KEA Algorithm Specifications, 1998.

[NIS99]   NIST FIPS Publication 46-3: Data Encryption Standard, 1999.

[NIS01a]   NIST FIPS Publication 197: Advanced Encryption Standard, 2001.

[NIS01b]   NIST FIPS Publication 800-22: A Statistical Test Suite for Random and Pseudorandom Number Generators for Cryptographic Applications, 2001.

[NIS02]   NIST FIPS Publication 180-2: Secure Hash Standard, 2002.

[NSN05]   L. Nguyen and R. Safavi-Naini. Dynamic k-Times Anonymous Authentication. In J. Ioannidis, A. D. Keromytis, and M. Yung, editors, *Proceeding of Applied Cryptography and Network Security - ACNS 2005*, volume 3531 of *Lecture Notes in Computer Science*, pages 318–333. Springer, 2005.

[Obe06]   G. Oberlecher. Anonyme Authentifizierungsverfahren. Master's thesis, Universität Klagenfurt, 2006.

[OP01]   T. Okamoto and D. Pointcheval. The Gap-Problems: A New Class of Problems for the Security of Cryptographic Schemes. In K. Kim, editor, *Proceedings of Public Key Cryptography - PKC'01*, volume 1992 of *Lecture Notes in Computer Science*, pages 104–118. Springer, 2001.

[OY91]   R. Ostrovsky and M. Yung. How to Withstand Mobile Virus Attacks. In *Proceedings of the 10th Annual ACM Symposium on Principles of Distributed Computing - PODC'91*, pages 51–59. ACM Press, 1991.

[Pai99]   P. Paillier. Public-Key Cryptosystems Based on Composite Degree Residuosity Classes. In J. Stern, editor, *Advances in Cryptology - EURO-CRYPT'99*, volume 1592 of *Lecture Notes in Computer Science*, pages 223–238. Springer, 1999.

[Ped91a]   T. P. Pedersen. A Threshold Cryptosystem without a Trusted Party (Extended Abstract). In D. W. Davies, editor, *Advances in Cryptology - EUROCRYPT'91*, volume 547 of *Lecture Notes in Computer Science*, pages 522–526. Springer, 1991.

[Ped91b]   T. P. Pedersen. Non-Interactive and Information-Theoretic Secure Verifiable Secret Sharing. In J. Feigenbaum, editor, *Advances in Cryptology - CRYPTO'91*, volume 576 of *Lecture Notes in Computer Science*, pages 129–140. Springer, 1991.

[Per]   Perl. http://perl.apache.org/docs/2.0/api/APR/UUID.html.

[PH78]   S. Pohlig and M. Hellman. An Improved Algorithm for Computing Logarithms over GF(p) and its Cryptographic Significance. *IEEE Transactions on Information Theory*, 24:106–110, 1978.

[PHP]   PHP. http://pecl.php.net/package/uuid.

[PHS03]   J. Pieprzyk, T. Hardjono, and J. Seberry. *Fundamentals of Computer Security*. Springer, 1 edition, 2003.

[PK01]   A. Pfitzmann and M. Köhntopp. Anonymity, Unobservability, and Pseudonymity – A Proposal for Terminology. In H. Federrath, editor, *Proceedings of Workshop on Design Issues in Anonymity and Unobservability*, volume 2009 of *Lecture Notes in Computer Science*, pages 1–9. Springer, 2001.

[PM04]   A. Pashalidis and C. J. Mitchell. A Security Model for Anonymous Credential Systems. In Y. Deswarte, F. Cuppens, S. Jajodia, and L. Wang, editors, *Proceedings of IFIP TC11 19th International Information Security Workshops*, pages 183–198. Kluwer, 2004.

[Pol78]   J. M. Pollard. Monte Carlo Methods for Index Computation (mod p). In *Mathematics of Computation*, volume 32, pages 918–924, 1978.

[Pol00]   J. M. Pollard. Kangaroos, Monopoly and Discrete Logarithms. *Journal of Cryptology*, 13(4):437–447, 2000.

[PS96]   D. Pointcheval and J. Stern. Security Proofs for Signature Schemes. In U. M. Maurer, editor, *Advances in Cryptology – EUROCRYPT'96*, volume 1070 of *Lecture Notes in Computer Science*, pages 387–398. Springer, 1996.

[PS00]   B. Pfitzmann and A.-R. Sadeghi. Anonymous Fingerprinting with Direct Non-repudiation. In T. Okamoto, editor, *Advances in Cryptology – ASIACRYPT'00*, volume 1976 of *Lecture Notes in Computer Science*, pages 401–414. Springer, 2000.

[PT05]   Regulierungsbehörde für Telekommunikation und Post. Bekanntmachung zur elektronischen Signatur nach dem Signaturgesetz und der Signaturverordnung. Bundesanzeiger Nr. 59, p. 4695–4696, January 2005.

[Pyt]   Python. http://docs.python.org/lib/module-uuid.html.

[Rab79]   M. O. Rabin. Digitalized Signatures and Public-Key Functions as Intractable as Factorization. Technical Report TR-212, Massachusetts Institute of Technology, Cambridge, MA, USA, 1979.

[Rab98]   T. Rabin. A Simplified Approach to Threshold and Proactive RSA. In H. Krawczyk, editor, *Advances in Cryptology – CRYPTO'98*, volume 1462 of *Lecture Notes in Computer Science*, pages 89–104. Springer, 1998.

[RE03]   W. Rankl and P. Effing. *Smart Card Handbook*. John Wiley & Sons, 2003.

[Res01]   P. Resnick. RFC2822: Internet Message Format. http://www.ietf.org/rfc/rfc2822.txt, 2001.

[Ric02]   J. Richter. *Applied Microsoft .Net Framework Programming*. Microsoft Press, 2002.

[Riv05]   R. Rivest. RFC1321: The MD5 Message-Digest Algorithm. http://www.ietf.org/rfc/rfc1321.txt, 2005.

[RS06]   S. Rass and M. Schaffer. Cryptographic Applications of Quadratic Imaginary Fields. Techn. Rep. TR-syssec-06-03, Klagenfurt University, 2006.

[RSA78]   R. L. Rivest, A. Shamir, and L. Adleman. A Method for Obtaining Digital Signatures and Public-Key Cryptosystems. *Communications of the ACM*, 21(2):120–126, 1978.

[RSG96]   M. G. Reed, P. F. Syverson, and D. M. Goldschlag. Proxies For Anonymous Routing. In *Proc. of the 12th Annual Computer Security Applications Conference – ACSAC'96*, pages 95–104. IEEE Comp. Society, 1996.

[RSS06a]   S. Rass, M. Schaffer, and P. Schartner. Anonymes Digitales Spielen im Casino. In P. Horster, editor, *DACH Mobility 2006*, pages 313–328. IT-Verlag, 2006.

[RSS06b]   S. Rass, M. Schaffer, and P. Schartner. Shared Generation of System-Wide Unique Keys for Discrete-Log Based Cryptosystems. Technical Report TR-syssec-06-02, Klagenfurt University, 2006.

[Rub]   Ruby. http://raa.ruby-lang.org/project/ruby-uuid/.

[Sal90]   Arto Salomaa. *Public-key cryptography*. Springer, 1990.

[Sch07]   M. Schaffer. Collision-Free Number Generation: Efficient Constructions, Privacy Issues, and Cryptographic Aspects. PhD-Thesis, Klagenfurt University, 2007.

[Sch89]   C. P. Schnorr. Efficient Identification and Signatures for Smart Cards. In G. Brassard, editor, *Advances in Cryptology – CRYPTO'89*, volume 435 of *Lecture Notes in Computer Science*, pages 239–252. Springer, 1989.

[Sch91]   C. P. Schnorr. Efficient Signature Generation by Smart Cards. *Journal of Cryptology*, 4(3):161–174, 1991.

[Sch96]   B. Schneier. *Applied Cryptography*. John Wiley & Sons, 1996.

[Sch01]   P. Schartner. *Security Tokens*. IT-Verlag, 2001.

[SG98]   V. Shoup and R. Gennaro. Securing Threshold Cryptosystems against
           Chosen Ciphertext Attack. In K. Nyberg, editor, *Advances in Cryptology
           – EUROCRYPT'98*, volume 1403 of *Lecture Notes in Computer Science*,
           pages 1–16. Springer, 1998.

[Sha49]   C. Shannon. Communication Theory of Secrecy Systems. *Bell System
           Technical Journal*, 28:656–715, 1949.

[Sha79]   A. Shamir. How to Share a Secret. *Communications of the ACM*,
           22(11):612–613, 1979.

[Sho97]   V. Shoup. Lower Bounds for Discrete Logarithms and Related Problems.
           In W. Fumy, editor, *Advances in Cryptology – EUROCRYPT'97*, volume
           1233 of *Lecture Notes in Computer Science*, pages 256–266. Springer, 1997.

[Sho98]   V. Shoup. Why Chosen Ciphertext Security Matters. Research Report
           RZ 3076 (#93122), IBM Research Division Zurich Research Laboratory,
           8830 Rüschlikon, Switzerland, November 1998.

[Sho00]   V. Shoup. Practical Threshold Signatures. In B. Preneel, editor, *Ad-
           vances in Cryptology – EUROCRYPT'00*, volume 1807 of *Lecture Notes
           in Computer Science*, pages 207–220. Springer, 2000.

[Sim83]   G. J. Simmons. A "weak" Privacy Protocol Using the RSA Crypto Algo-
           rithm. *Cryptologia*, 7:180–182, 1983.

[Sim91]   G. J. Simmons. Geometric Shared Secret and/or Shared Control Schemes.
           In A. Menezes and S. A. Vanstone, editors, *Advances in Cryptology –
           CRYPTO'90*, volume 537 of *Lecture Notes in Computer Science*, pages
           216–241. Springer, 1991.

[SR07]   M. Schaffer and S. Rass. The Fusion Discrete Logarithm Problem Family.
           Technical Report TR-syssec-07-02, Klagenfurt University, 2007.

[SR08]   M. Schaffer and S. Rass. Secure Collision-Free Distributed Key Genera-
           tion for Discrete-Logarithm-Based Threshold Cryptosystems. In G. Dor-
           fer, G. Eigenthaler, H. Kautschitsch, W. More and W. B. Müller, editors,
           *Proceedings of the Klagenfurt Workshop on General Algebra ("73. Arbeit-
           stagung Allgemeine Algebra")*, pages 159–174. Johannes Heyn, 2008.

[SS01]   S. G. Stubblebine and P. F. Syverson. Authentic Attributes with Fine-Grained Anonymity Protection. In *Proceedings of Financial Cryptography – FC'00*, pages 276–294. Springer, 2001.

[SS05]   P. Schartner and M. Schaffer. Unique User-Generated Digital Pseudonyms. In V. Gorodetsky, I. V. Kotenko, and V. A. Skormin, editors, *Proceedings of Mathematical Methods, Models, and Architectures for Computer Network Security – MMM-ACNS 2005*, volume 3685 of *Lecture Notes in Computer Science*, pages 194–205. Springer, 2005.

[SS06]   M. Schaffer and P. Schartner. Anonymous Authentication with Optional Shared Anonymity Revocation and Linkability. In J. Domingo-Ferrer, J. Posegga, and D. Schreckling, editors, *Proceedings of Smart Card Research and Advanced Applications – CARDIS'06*, volume 3928 of *Lecture Notes in Computer Science*, pages 206–221. Springer, 2006.

[SS07]   M. Schaffer and P. Schartner. Implementing Collision-Free Number Generators on JavaCards. Technical Report TR-syssec-07-03, Klagenfurt University, 2007.

[SS08]   P. Schartner and M. Schaffer. Protecting Privacy in Medical Databases. In *Proceedings of the International Conference on Health Informatics (HEALTH'08)*, volume I, pages 51–58, ISBN 978-989-8111-16-6, 2008.

[SSR07a] M. Schaffer, P. Schartner, and S. Rass. Efficient Generation of Unique Numbers for Secure Applications. Technical Report TR-syssec-07-01, Klagenfurt University, 2007.

[SSR07b] M. Schaffer, P. Schartner, and S. Rass. Universally Unique Identifiers: How to ensure Uniqueness while Protecting the Issuer's Privacy. In S. Alissi and H. R. Arabnia, editors, *Proceedings of the 2007 International Conference on Security & Management – SAM'07*, pages 198–204. CSREA Press, 2007.

[Sta96]   M. Stadler. Publicly Verifiable Secret Sharing. In U. M. Maurer, editor, *Advances in Cryptology – EUROCRYPT'96*, volume 1070 of *Lecture Notes in Computer Science*, pages 190–199. Springer, 1996.

[SV97]   A. Sahai and S. Vadhan. A Complete Promise Problem for Statistical Zero-Knowledge. In *Proc. of the 38th Annual IEEE Symp. on Foundations of Computer Science – FOCS'97*, pages 448–457. IEEE Press, 1997.

[SW78]   O. Schafmeister and H. Wiebe. *Grundzüge der Algebra*. Teubner, 1978.

[SYT05]   K. Sako, S. Yonezawa, and I. Teranishi. Anonymous Authentication: For
            Privacy and Security. *NEC J. of Advanced Technology*, 2(1):79–83, 2005.

[Tak98]   T. Takagi. Fast RSA-Type Cryptosystem Modulo $p^k q$. In H. Krawczyk,
            editor, *Advances in Cryptology – CRYPTO'98*, volume 1462 of *Lecture
            Notes in Computer Science*, pages 318–326. Springer, 1998.

[Tan03]   A. S. Tanenbaum. *Computer Networks*. Prentice Hall, 4th edition, 2003.

[Tes01]   E. Teske. Square-Root Algorithms for the Discrete Logarithm Problem (A
            Survey). In *Proceedings of Public-Key Cryptography and Computational
            Number Theory*, pages 283–301. W. de Gruyter, 2001.

[TFS04]   I. Teranishi, J. Furukawa, and K. Sako. k-Times Anonymous Authenti-
            cation (Extended Abstract). In P. J. Lee, editor, *Advances in Cryptology
            – ASIACRYPT'04*, volume 3329 of *Lecture Notes in Computer Science*,
            pages 308–322. Springer, 2004.

[Tlc]   Tlc. `http://tcllib.sourceforge.net/doc/uuid.html`.

[Uni06]   United States Department of Health and Human Services, Office for Civil
            Rights. HIPAA Administrative Simplification – Regulation Text. 45 CFR
            Parts 160, 162, and 164, 2006.

[Wal98]   I. Walker. The Economic Analysis of Lotteries. *Economic Policy*,
            13(27):357–402, 1998.

[Wol99]   S. Wolf. *Information-Theoretically and Computationally Secure Key
            Agreement in Cryptography*. ETH Dissertation No. 13138, Swiss Federal
            Institute of Technology, ETH Zurich, 1999.

[WWW02]   T. Wong, C. Wang, and J. Wing. Verifiable Secret Redistribution for
            Threshold Sharing Schemes. Technical Report CMU-CS-02-114-R, School
            of Computer Science, Carnegie Mellon University, Pittsburgh, 2002.

[XML]   XML. `http://www.w3.org/TR/1998/NOTE-XML-data-0105/`.

[Yao82a]   A. C. Yao. Protocols for Secure Computations. In *Proceedings of the
            23rd Annual IEEE Symposium on Foundations of Computer Science –
            FOCS'82*, pages 160–164. IEEE Press, 1982.

[Yao82b]   A. C. Yao. Theory and Applications of Trapdoor Functions. In *Proceedings
            of the 23rd Annual IEEE Symposium on Foundations of Computer Science
            – FOCS'82*, pages 80–91. IEEE Press, 1982.

www.ingramcontent.com/pod-product-compliance
Lightning Source LLC
LaVergne TN
LVHW022309060326
832902LV00020B/3366